Encyclopaedia of Environmental Pollution and Awareness in 21st Century Series

WORLD ENVIRONMENT

Edited by
V.K. Prabhakar

ANMOL PUBLICATIONS PVT. LTD.
NEW DELHI - 110 002 (INDIA)

ANMOL PUBLICATIONS PVT. LTD.
4374/4B, Ansari Road, Daryaganj
New Delhi - 110 002

World Environment

First Edition, 2001

ISBN 81-261-0935-1

PRINTED IN INDIA

Published by J.L. Kumar for Anmol Publications Pvt. Ltd., New Delhi - 110 002 and Printed at Mehra Offset Press, Delhi.

Contents

PREFACE

Today, the cry of 'pollution' is heard from all the nooks and corners of the world, and pollution has become a major threat to the very existence of mankind on this earth. It is the major challenge of our times.

The pollution of various resources has gone to such an extent that we are unable to breathe fresh air and drink fresh water. On one hand the advancements of science and technology have added to the human comforts by giving us automobiles, electrical appliances, supersonic jets, space crafts, better medicines, better chemicals to control harmful insects and other pests, etc., but, on the other hand, they have given us a very serious problem to face that is pollution. Pollution can be defined as an undesirable change in the physical, chemical, or biological characteristics of the air, water, or land that can harmfully affect health, survival, or activities of humans or other living organisms.

Pollution does not have to cause physical harm, but it may merely interfere with human activities. Chemicals spewed into the air or water from an industrial plant may be harmful to humans and other organisms living near the plant. If expensive pollution controls are required, the plant may be forced to shut down.

India faces serious problems of environmental pollution. Water pollution from untreated community and industrial wastes flowing into our rivers and streams, increasing use of fertilizers and pesticides, dumping of organic wastes etc., are issues that need to be tackled on the priority. Air pollution, long regarded as the bane of industrialised nations looms large over our country.

Here, we are presenting this *Encyclopaedia of Environmental Pollution and Awareness in Twenty-first Century* in fifty volumes with a view to make people conscious about the seriousness of pollution in the present time and its remedies.

Editor

1 NATURAL RESOURCES

Technology cannot bring back an extinct animal resource or a paved over wilderness area, but it can extend the supply of some resources by improving them, using them more efficiently, or recycling them. For example, a unit of today's steel provides 43% more structural support that it did a decade ago. Today we get 7 times more electrical power from 1 ton of coal than we did in 1900. Similarly the energy needed to produce 1 ton of pig iron has fallen eight fold since 1800. In the United States in 1900 only 10% of the copper was recycled. Today about 40% is recycled. However, while many matter resources such as copper lead and silver, can be recycled, we can never recycle energy resources.

Once a fossil fuel resource, such as coal, oil or natural gas, is burned, it is gone forever as a useful energy source. The concentrated energy in the fossil fuel is released as heat, which is eventually dispersed into the earth's atmosphere. From there it flows back into space. Sometimes technology can solve the problem of a scarce resource by finding a substitute or replacement. For instance as a structural material bronze replaced stone, iron replaced bronze, steel replaced iron and now aluminum and reinforced plastics are replacing steel for some structural uses. As an energy source, animals replaced human muscle power coal (to produce steam) replaced animals and oil replaced coal for many uses.

There is a vigorous debate over whether a combination of coal derived energy and nuclear power (obtained from Uranium) or a combination of solar wind and plant (bio mass) energy may soon replace petroleum and natural gas.

In addition to technology resource use is tied to economics. Something is useful as a resource only if it can be made available at a reasonable cost. For example once we deplete the easily available supplies of a resource, we have to look harder and dig deeper to find remaining supplies. In the costs of findings and making a scarce resource available rise, the resource will eventually become too expensive for most people. Higher costs may stimulate a new search for new supplies or make mining and processing lower grade deposits economically feasible. But regardless of what we are willing and able to pay, we can't get a resource out of earth if it is not there.

There can also be an economic limit to recycling. Typically recycling is cheaper than mining virgin materials but only if the material to be recycled is not too widely dispersed. For example, if products made from iron and steel, such as cars and toys are thrown away, buried and widely scattered through use labour and energy costs for finding and collecting the objects may be too expensive to make recycling feasible.

The continueity of a resource may also depend on the impact its mining, processing, and use has on the environment. Even if affordable supplies of a resource is available, its use (at least for certain purposes) may have to be abandoned if this use seriously threatens human and other forms of life. Sometimes the environmental effects of a resource use can be minimized and cleaned up. But this cost money and in some cases the cost may be so high that we can no longer afford to use the resource.

The sum of all physical, chemical, biological and social factors which compose the surroundings of man is referred to as environment and each element of these surroundings constitutes a resource on which man draws in order to develop a better life. Thus, any part of our natural environment—such as land, water, air minerals, forest, rangeland, wildlife, fish or even human population — that man can utilize to promote his welfare may be regarded as a natural resource. Ramade defined a resource as a form of energy and/or matter which is essential for the functioning of organisms, populations, and ecosystems. In the particular case of humans, a resource in his words is any form of energy or matter essential for the fulfilment of physiological, socio-economic, and cultural needs, both at the individual level and that of the community.

The functioning of ecosystem or of technological civilization involves a continuous flow and consumption of energy sun being the source of it as regards ecosystem and fuels as regards technological civilization. In the light of this fact, a resource may be defined as one of the various forms of primary energy present in the nature.

The five basic ecological variables — energy, matter, space, time and diversity are sometimes combinedly called natural resources. As such, laws which govern changes in these quantities have great applicability in the resource use.

The Classic subdivisions of resources include renewable and non-renewable resources. Renewable resources are those which are reproducible and are obtained from the biomass of living organism, e.g. forests and fish. Resources which are not reproducible and are obtained from the finite non-living reserves are called non-renewable resources e.g. coal and metals. Solar energy although having a finite life, as a special case, is considered as a renewable resource in an much as solar 'stocks' are inexhaustible on the human scale.

Some authors prefer to classify resources, into biotic (or living) and abiotic (or non-living) resources, forest, agriculture, fish and wildlife falling in the first category and land, water minerals etc. in the second.

Quantity, mutability and reusability are the attributes on which this classification is based. Inexhaustible and exhaustible are main categories of resources based on their stocks in nature. Inexhaustible resources are further divided into two groups. Immutable and mutable resources, depending upon the possibility of their qualitative degradation as a result of man's activity. On the other hand, exhaustible resources, being limited in occurrence, are vulnerable to both quantitative and qualitative degradation. Thus their availability depends upon the method of use, so these are classed into maintainable and non-maintainable resources. Supply of maintainable resources could be made to last for long through wise use, however, some of them have capacity to reproduce called renewable, while other cannot reproduce-called non-renewable resources. Non-maintainable resources have static supply and when destroyed or consumptively used are irreplaceable, yet some of them have potentiality for reuse-called reusable resources, and rest whose permanent exhaustion certain are called non-reusable resources.

The term "conservation" to many means "hoarding" or to control supply of goods in such a way that some part is left for the nature. Most people interpret conservation as total protection or restriction in consumptive use of resources, therefore, wildlife conservation to them means a total ban on killing of any animal and the forest conservation means a complete moratorium on all feelings the term, in fact, is derived from two Latin words-con (together) and servare (to keep or guard), thus meaning 'to keep together'. The dictionary meaning of the word is an act of preservation. The term was proposed by Giffor Pinchot in 1988, who conceived it from the title of 'conservares' given to British officials in India appointed to look after the protection of natural resources.

Prof. H.M. Rose defined conservation as "the optimum allocation of natural, human and cultural resources in the scheme of national development, where by maximum economic and social security will be assured. In 1967 the American president John F. Kennedy looked upon conservation as "the wish use of our natural environment... The prevention of waste and despoilment while preserving, improving and renewing and the quality and usefulness of all our resource" World Conservation Strategy declines conservation as "the management of human use of the biosphere so that it may yield the greatest sustainable benefits to present generation, while maintaining its potential to the needs and aspirations of future generations" thus the conservation is a practice embracing preservation maintenance, sustainable utilization, restoration and enhancement of the natural environment. Summarily, thus the aim of conservation is two fold: (i) preservation of quality of environment, and (ii) to ensure continuous yield of useful material-living or non-living-by establishing a balanced cycle of harvest and renewal.

The concept of resource conservation has now come under a still wider field on environmental management which is, in essence, the process of allocating natural and man made resources so as to make optimum use of the environment in satisfying basic human needs at the minimum and more, as far as possible, for an indefinite future, and at the same time preventing depletion and degradation of resources.

Environmental management being a too vague, general, or trivial subject has not developed as a full-fledged discipline so far, further, in many developing countries it is viewed as being irrelevant and counter productive, while in advanced industrialized countries it is often interpreted as a means of coping with the 'burden' of environmental legislation and problems put forth by 'over-zealous' environmentalists what to talk of under developed world! The global population, however cannot and should not overlook and deny the last deterioration of environmental world over resulting in dwindling natural resources, pollution of air, water and land, extermination of a number of life forms, down going healths status, danger to cultural assets and many socio-economic backlashes. The root of the problem lies in too may people using too many resources wastefully.

Since the total natural resources available are limited, the way out to the problem seems only through control of population and also the per capital consumption of resources. Because it is difficult to bring down per capital consumption, the recycling and stringent conservation can bring hopes for sustained development without reducing per capital use of resources.

Land

Soil consists of the weathered layer of the earth's crust with living organisms and products of their decay intermingled. Lying on the rocky core of the earth, this complex mixture is one of the four prime requisites for life. Along with sunlight, air and water, soil supports all living beings on land, plants as well as animal. Altogether terrestrial plants have their roots in soil from which they absorb life sustaining water and nutrients. Man and all land animals, in turn, depend on these plants for food directly or indirectly. Foods, vegetable oils and fats, forage for livestock, fiber, wood and various other indispensable agricultural products come from soil. Thus, a nations economic well being is inextricably linked with the fertility and abundance of her soil resource because of mounting demands of swelling population and long mismanagement would put in jeopardy the very survival of man.

Land use and abuse

The land although appears to be an unlimited resource but, in fact, its injudicious use would limit the availability of this indispensable life-support system. Thus, a land use may be poor or good, if in a region of low rainfall grasslands are replaced by wheat crops, a temporary desert will eventually result; or if the human habitation and factories are developed on flood plains, the damage is almost inevitable. These are the instances of poor land use. Conversely, if the grasslands are used for moderate grazing by livestock, and it the flood plains are used for forestry or agriculture, these will imply good land use. Ecological principles have great applicability in land use planning. Classification of use should, therefore, be based on natural ecological features of the land in question viz, soil type, slope, biotic communities etc.

Rural land-use Planning

To a farmer, soil is the essence of survival and his economic well being is firmly linked with the fertility of his land. In rural land-use planning, concentration is chiefly devoted to creating and developing more farmland by removing forests an natural grasslands, drawing marshes and swamps, channeling streams for irrigation, and so on. Unfortunately, no effort is made to save existing prime farmland from degradation by ill-planned development. Rather than exclusive farm, range or forestry oriented planning, the land use should be made under broad consideration of social, economic, and ecological intricacies.

Urban land-use planning

Deterioration quality of urban and suburban environment is to a great extent the result of injudicious land use and is a threat to the whole socio-economic system. Thus planned cities are as necessary as planned farms. Although, the implementation of land-use plan is infinitely more difficult in urban areas than in rural ones because of the human problems involved, yet if not heeded it would lead to chronic pollution, epidemic, and social disorder. In any urban land-use planning it should be seen that: (i) at least a third of all land remains as protective open space in the form of parks, green belts, or refuges etc. including adequate space for efficient treatment of domestic and industrial wastes: (ii) residential housing around villages or cities should be developed in clusters with each unit separated by broad green belts; (iii) typical topographic situation like stream valleys steep slopes and marshes etc. should be retained free from houses and other high-density uses; and (iv) waste disposal areas and industries should be kept apart from human habitations.

Management of Land

From the above account it is clear that major problems emerging out of abuse of land in cities include unsafe and unhygienic dwelling and those in rural areas are of erosion of soil and loss of fertility. While urban problems can be prevented and cured by proper town planning and municipal facilities, the problems relating to rural areas needs long term soil conservation programme.

Prevention and control of soil erosion: The primary purpose of soil conservation is to prevent soil erosion and heal the damage where it has not advanced too far to respond to curative methods. This is possible largely through two means:

1. The land should wear a vegetational cover throughout the year. Plant roots bind the top soil and prevent it from being drifted so easily with water or air. The vegetative cover also slows up runoff water and causes much of it to percolate into ground that promotes more vegetative growth. The vegetation-whether grasses, trees, or cultivated crops-should fit to the capabilities of the soil and water available.
2. Engineering and agronomic practices should be applied conjointly to counteract the erosive force of both water and wind on the soil. These include contour cultivation, cover crops and crop rotation, and contour strip cropping, used alone or in combination with terracing where it is required. When terraces are properly constructed and adequately supported by approved cropping and tillage practices, they prove to be one of the most effective erosion-control measures applicable to cultivated sloping lands.

Reclamation of eroded lands: Conservationists are divided into two camps over the issue of reclamation of land already arrested by ravines. One camp of conservationists and naturalists believes in the policy of "forgive and forget" and advocates that ravines should be allowed to exist, while the other feels that efforts should be made to regain the lost land through buldozing or some other means. If the ravines are left in there present state, these will spread further and engulf more of productive land. While the violent means to regain the ravines in their near original condition may provoke numerous agencies of nature to take revenge on man. Thus, following considerations are important in negotiating the eroded land:

i. Ravines should be provided with sufficient and suitable vegetative cover.
ii. Instead of agriculture, these lands should be reclaimed for forestry, pasture, or horticulture.
iii. Their deficiency in nutrients and moisture for plant growth should be improved.
iv. Further misuse of such land should be prevented. Over-trampling by man and cattle should be avoided.
v. Past damages should be repaired and recompensed under strict ecological consideration. For instance depressions may be filled up, slopes easend out, cattle, trials fenced etc.
vi. Vegetative cover provided should be protected against reckless destruction by local population.

Water

The origin of water on the earth is not clear so far. However, the current presumption is that the primordial earth had no oceans, and perhaps very little atmosphere. It is believed that the volatile constituents bound in the earth's crust, oozing the surface through volcanos, rock movements, and hot springs, condensed to form the ocean and the atmosphere. This way, perhaps this remarkable combination to hydrogen and oxygen — called water, came into being and eventually became an indispendable component of the earth's environment.

Water moves from the ocean to air to land to ocean in set cyclic pattern called hydrologic cycle. In other words, this cycle represents a continuous process of exchange of water from the earth's surface to the atmosphere and vice versa under the influence of solar heat.

It is estimated that of the total available water 97.3 per cent is contained in the oceans and out of remaining

2.7 per cent most is in solid form. The amount of water actually available in the atmosphere is a very minute fraction and estimated to be only 1×10^{-5} of the world's total water.

The factors which control the process of evaporation and evapotranspiration include temperature, humidity, and wind. It is a continuous process going on day and night from the ocean surface, ground surface inland waters (like lakes and streams), plant and animal surfaces. Rising in the atmosphere this water forms clouds that float around in gaseous or droplet form. The atmosphere above the earth up to varying heights of 10 to 17 km (depending upon latitude and season) is called the troposphere up to which the clouds float and in which all weather changes take place. Water returns to the earth in the form of precipitation, most of which takes place over the ocean and a little on the land surface. Rainfall on the land surface may have following fates: it may percolate through the soil and become a part of ground water regime; accumulate in a pond, lake or reservoir as inland surface water; flow down to the ocean in the form of stream; taken up by plants and animals; or evaporate.

Water Requirements and uses

Goethe rightly said that, "Every thing originated in the water, and everything is sustained by water" Water is needed to fulfill diverse requirements in so many diverse ways. It is vital to life since for all physiological activities of plants and is essential for animals as well. On an average it constitutes 80 percent of protoplasm ranging from 8 per cent in dry dormant seeds to 95 per cent jelly fish. It serves not only to quench the thirst but also to meet the food requirements because it is an essential raw material in the process of photosynthesis through which green plants make food that is used by all trophic levels directly or indirectly. Ambient humidity which is indeed the invisible form of water that surrounds us is necessary to prevent desiccation of terrestrial life forms. Thus, it can be said that water is a biologically essential and it serves as milieu internal as well as milieu internal.

Flood plains of rivers have been the cradle of civilization and centres of population since ancient times. The reason being their high fertility an easy accessibility to water for various uses like drinking, washing and transport. Egypt was developed on the flood plains of the Nile: Babylon on the flood plain of the Tigris: and North Indian on the flood plaims of the Ganges. Even today the Mississippi plain is about the most densely populated area in the USA as flood plains of the Rhine in Europe, and the Yellow river in China.

Besides temperature, water is the other such key factor that influences the global ecology. Covering about 70 per cent of land surface and circulating in hydrologic cycle water influences weather and climate on any region and thus its flora and fauna.

Agriculture is dependent on water, may it be rainfall, or surface or ground water irrigation. It eventually is a more important commodity in a country like India because her largely agrarian economy depends critically on water. Huge amount of protein rich food in the form of fish shell fish, and prawn is obtained from the inland as well as very potent industry.

The kinetic energy of gravitational flow of water is tapped and transformed into electricity, what is called hydro power. This concept of electricity generation is one of the oldest and cheapest tricks largely used world over. Moreover, it leaves no pollution burden on nature as in case of thermal or nuclear power generation.

Hardly any industry can do without water. It has

multifarious application in industrial processes viz, a raw material, solvent, chemical reactant, coolant, and cleaning agent.

"Water, water every where not any drop to drink is the situation not only found amidst sea but also at places where industrial and domestic refuse make the benevolent water sources so severely contaminated with pathogens and or toxins that they are lift impotable. Being the best dilution medium it is frequently used as the cheapest and convenient way for liquid and solid waste disposal. Further is was practically impossible to maintain sanitary conditions without water for it is needed in all common cleaning and washing practices.

Water at many places serves as a very convenient means of navigation. Besides, it adds to the aesthetic value of the landscape and provides opportunities for recreation and sport.

Water being unevenly distributed over the land, influences interstate and international relations and thus is the cause of what is known as hydro-politics.

It can, therefore, be said that while water is essential for life it is significantly linked with social, economic, political and ecological intricacies.

Problems and Management of Water

Management of water implies making the best use of available water resources for human benefit while not only preventing and controlling its depletion and degradation but also developing it in view of the present and future needs. Water, like forest, is a multipurpose resource and it is important to see that its various uses should not conflict with each other and it can be enjoyed in its totality by man and others. Thus, its right allocation, and quantitative and qualitative conservation are the primary tasks before water managers.

Floods, droughts, improper use, pollution and disease transmission are the important problems related to water. Because usable inland water resources constitute a very minute fraction of the world's total water and because man largely depends upon this minute fraction of water for his various activities, it is important to make efficient use of available water and at the same time prevent abuse of it. This has to be considered not only in giant irrigation projects or industrial and municipal processes but also in every day life. Taps running unpurposefully, leaking water supply pipes, using filtered water for gardens and building construction are a few instances of inefficient and misuse of water in daily life.

Water being the best and convenient dilution medium is the worst affected resource by pollution. To a common man pollution of water means only the drainage of industrial effluents or domestic sewage in natural waters. It may astonish a lay reader that even pure distilled water when released in sea may act as a pollutant since it alters the normal salinity of sea water which adversely affects the marine life. Besides chemical alterations, changes in physical condition of water also have grave effects on the biota of system as during thermal pollution caused by the discharge of heated effluents in a body of water. Pathogens, chiefly entering natural waters through municipal wastes rich in excreta and other organic filth, also render the water unsuitable for human use.

In fact, it is not only the man who has to use the water resources by these are equally important to land animals and more to those fish and other aquatic animals who live in it. Thus pollution of a body of water does prove detrimental not only to the human society but also to regional ecology and the ecology of very aquatic ecosystem. It is, therefore, essential that the water should not be treated as a simple repository of waste disposal, and if inevitable, the wastes only after proper treatment should be released in it under strict ecological consideration.

The qualitative degradation of water is also in a way quantitative depletion of usable resources. Therefore, recycling of waste water after due treatment would relieve the water scarcity to a great extent, especially in regions marred by water shortage.

Future needs and alternate sources of water

The demand for usable water like any other resource, is increasing fast with the continuous increase in world population and continuous increase in per capital demand linked with more of sewage disposal and transport prolific use of water-using gadgets (like air cooling and conditioning devices, washing machines etc.) and increased recreational activities. According to an estimate, the daily per capital demand for water of an American was 600 gallons in 1900 which increased to 1,500 gallons in 1960 and is expected to be increased upto 2,700 gallons by 2000. The fact is no different in developing world. Thus, the question is where will the water needed for fulfilling increasing demands come from? The answer is not so simple and can only be found through long term planning. Possible means of meeting out the in pending water deficit problem include: (i) reclamation of sewage and waste water, (ii) development of ground water sources and surface storages; (iii) long range forecasting of rain; (iv) rain making; (v) transfer of sulfur water, and (vi) desalination of sea water.

Agriculture

The practice through which specific plant species are cared and managed so as to obtain maximum yield of consumable parts of the plant is called agriculture. An agricultural field is, in fact, a cropland ecosystem in which ecological principles involving limiting factors and productivity provide great applicability.

The demand for food production , linked with tremendous population growth, has been increasing fast world over. This has been possible to feed increasing number of mouth through increased agricultural production what is called green revolution. It has resulted chiefly from the industrialization of agriculture, which involves highly domestic plant species, sophisticated chemicals, and large fuel energy subsidies. According to an estimate, to double crop yield through industrialized agriculture, a ten fold increase in fertilizers, pesticides and horse power is required.

The green revolution through industrialized agriculture (fuel powered) as compared to conventional agriculture (man-and domesticated-animal powered) has following backlashes:

i. Such an agriculture involving huge amounts of fuel, and chemicals in the form of fertilizers, pesticides and weedicides, is the chief cause of environmental pollution.
ii. It may lead to social backlashes when a quick shift from small farms to large industrialized farms takes place. In this process small farmers and agriculture workers are driven off the land to cities where they may not get work, food and housing—a situation of despair and social breakdown.

In retrospect, strong measures should be taken to control the population growth and to improve the conventional agricultural practice with application of current agricultural research under due social and environmental consideration. For a greater and sustained agricultural production following considerations are important:

i. Available land acreage should be properly and judiciously utilized.
ii. Soil fertility should be increased through wise use of fertilizers and organic manures.
iii. High yielding and disease resistant plant varieties should be introduced.
iv. Mixed cropping should be practicised wherever possible.
v. Integrated and balanced use of available water sources, surface and ground water should be made.
vi. Weeds and pests should be efficiently controlled. Integrated pest control practices should be preferred over total reliance on chemical pesticides.
vii. Soil erosion and loss of nutrients should be prevented through maintaining vegetation cover throughout the year.
viii. Exotic varieties should be introduced only after due consideration of long term impacts.

Forestry

A forest is a natural ecosystem having multi species and multi aged trees as dominant community. Forests cover about one-third of the earth's land surface of which about 50 per cent is occupied by tropical forest. Of the total geographical area of India 22.74 per cent is forest as against a minimum of 33 per cent forest cover prescribed under National Forest Policy.

Importance of forests

Functions of a forest may broadly be classified into following three categories:

i. Protective functions: These include the protective role of forests against soil erosion, droughts, floods intense radiation etc.

ii. Productive functions: Forests are the source of wood and many other products like gums, resins, fibers, medicines, katha, honey, bidiwrappers, pulp, paper etc.

iii. Accessory functions: These include the role of forests in recreation, aesthetics, and as habitat of diverge wildlife etc.

Thus, it can be said that forests are important in two ways—ecologically and economically.

Ecological significance: Environmentally, the forests are very important. They help in balancing oxygen and carbon dioxide level in atmosphere, regulate earth's temperature regime and hydrologic cycle. Forests increase local precipitation and water holding capacity of soil, thus preventing drought situation. Vegetation cover provided by forest impedes the velocity of runoff on soil surface checks soil erosion, silting and landslides, thus reducing the danger of flood. The litter derived from fallen leaves maintains fertility of soil by returning the nutrients. Forests also act as refuge of wild animals and provided protection to them against strong, cold or hot and dry winds, solar radiation, rain and enemies.

Economic significance: Wood which has various applications in domestic and industrial processes, is the chief product of forest. Wood when used as fuel, has certain advantages over coal as its sulphur and ash contents are very low but, at the same times, excessive use of fuel wood mens pressure on forests which have many functions to serve. Wood may also be converted into solid (coke) liquid and gaseous fuels. Timber is an important material in building construction and day-to-day uses. Forest provides raw material for various wood-based industries viz. pulp and paper, composite wood, rayon and other man made fibers, sport goods, furniture, boat building matches etc. Miscellaneous products, like bamboos, resins, gums, some oils, fibers, medicines, katha, lack, shellac etc., are also obtained from forests.

Depletion of forests

Increased urbanization, industrialization and mining has entailed indiscriminate felling, of trees has denudation of forests. Nobody will, however, ever learn that who set the first assault on which tree in the history, but this practice of logging down trees for human use has been going on or centuries unabated. Colonists used logs for fire wood and making huts. They also made handles for their tools, furniture and fence posts etc. from it. For them forest had only a present and a past-no future; thus they resorted to ruthless and accelerated exploitation of available forest—a story that still continues with more serious backlashes earth has ever witnessed. India alone is losing more than 1.5 million hectare of forest cover each year and according to Tiwari (1982) 22 million hectare of forests have been destroyed during the last three decades due to over exploitation and misuse and conversion to agricultural fields. The depleted forest wealth would simply deprive the man of economic and environmental values offered by forest, as discussed earlier.

Management of forests

In forestry, the yield is chiefly obtained from the accumulated growth of the past, so this industry has to depend upon the annual growth. Forests are managed to obtain a sustained yield which is obviously possible through sustained growth. Following considerations are important in this direction.

i. A forest should have in correct proportion the trees of different ages and classes so that continuous removal of exploitable produce would not affect the future yield.

ii. Each tree crop should put on normal growth increments.
iii. No obstacle should exist to the continuous harvest of exploitable stems.
iv. Unrestrained felling should be prevented.

For all such planning and decision making, sound ecological knowledge is desirable that include knowledge about climatic conditions, of biotic community, successional trend, life cycle of important tree species methods of arresting growth of undesirable trees and promoting growth of high yield tree species, and biotic inter relationship (such weeds, competitions, pets, parasites and diseases etc.)

Of various tactics employed in forest management, same deserving special considerations are as follows: (i) restrained felling (ii) block and selective cutting (iii) reforestation (iv) control to disasters and (v) recycling of forest products.

Rangeland

A range is the grassland that provides natural pasture or grazing animals. Grasslands occur where rainfall is too low to support the forest community but is higher than that which results in desert community. Generally, in grassland precipitation ranges between 25 to 75 cm. However, depending upon edaphic factors, such as a high water table, grasslands may also occur in regions of forest climates. Soil of a grassland is characterized by having high humus content.

Over 6, 000 species of grasses, adapted for a great variety of climatic regimes, cover about 20 per cent of earth's surface. In size, they vary from as small plants as about 1 cm. to over 35 cm. tall bamboos.

Importance of Rangeland

Man has been cultivating various grass species as his food and as fodder for his domesticated livestock since time immemorial. History of human civilization the witness of dependence of is man on the seeds of various grasses as his own food, for instance, Indo-Chinese, American and Mediterranean cultures were based on ice, maize and wheat respectively while forage was used as fodder for domesticated livestock.

A range, if not serving as a direct source of food for man, still serves the mankind in a number of indirect ways. Solar energy transformed into food energy by the green cover of the range is available to primary consumers (grazers-domesticated or wild) as food, where in a part of it is converted into meat. Of considerable importance to; man are these herbivores and their products such as meat, milk, skins, hides, horns and wool etc. It can thus be said that at the root of all biological resources that, man utilizes is the forage.

Depletion and Degradation of Rangeland

Man would not destroy an artificial pasture developed by input of his hard labour and money than a natural pasture. Thus, like all natural resources the rangeland, inflicted with human stress is depleted and degraded. Cause of such stress may be ignorance, greed or incompetence of man. Phenomenal growth of cattle industry, on one hand, is responsible for ruthless over grazing. While on the other hand, wide expansion of urbanization, industrialization, and agriculture has resulted in encroachment of rangeland. Often, out of greed too a rancher allows a number of head to graze beyond the carrying capacity of the range being ignorant to the fact that too many sick and weak head would not fetch home as much value as a few head in prime condition of health after grazing on enough good forage.

Community or ecological indicators provide the most practical approach of determining whether or not a range is being utilized properly (Humphery, 1962), Range managers use the term "decreasers" for palatable species vulnerable to grazing their disappearing is an indicator of grazing stress. When such indicators are not heeded unpalatable "weeds". Once established, it may become hard to eradicate the weeds and not only this such vegetation will invite a variety of pest animals like rodents and insects. This makes clear the long-term consequences of overgrazing in a range.

There is increasing need to restore the depleted and degraded rangeland not only out of economic (livestock production) but also ecological consideration since rangeland depletion would also cause problems like soil erosion, spread of pests etc.

Range management

Stoddard & Smith (1955) defined the range management as " the science and art of planning and direction range use as to obtain the maximum livestock production consistent with conversion of range resources."

Efforts to manage land include: regulation of grazing pressure: manipulation of stock (herd) distribution: burning of range: reseeding of rangeland: selection of cattle breeds: and control of weeds, pests and predators.

Wildlife

Even on global level there appears to be a great deal of ambiguity attached to the term "wildlife". Webster's Dictionary defines wildlife as living things that are neither human nor domesticated, specially mammals, birds and fish hunted by man. In American usage, term 'wildlife' is used to designate the games and excludes fish, thus they have Fish & Wildlife Services at the federal level. In India too, the common understanding of the people regarding wildlife is of wild animals and shikar, more particularly mammals, birds and reptiles. However, a report entitled Wildlife Conservation in India' (1970) of the India Board for Wildlife states wildlife as the entire native uncultivated flora and fauna of the country. The Wildlife (Protection) Act, 1972, also defines wildlife as any animal, bees, butterflies, crustacea, firsh and moth, and aquatic and land vegetation which form part of any habitat.

Thus a comprehensive definition of wildlife may be given as-any or all non-cultivated and non-domesticated life including both, plant and animals.

Importance of Wildlife

The Wildlife resources provide aesthetic, recreational and economic benefits and at the same time are important ecologically. Nature's millions of forms of life on this earth perform some or the other definite biological function which keeps the delicate balance of our exosphere. In nature flora and fauna go hand-in-hand and one cannot exist without the other. Bird and insects are necessary for the cross pollination of flowers and these with other animals also serve for dispersal and propagation of vegetation. Not only this, the health of vegetation also depends on natural browsing and grazing. Needless to mention that animals, directly or indirectly depend on green plants for their food. Besides food plant also provide shelter to a variety of animal species equally important are the relations among animal populations. If too many herbivores are killed, the predators, depriced of their natural food, will leave forest, become vermin, and prey upon domestic livestock or even man. On the other hand if too may carnivores are killed the herbivores because of release of predation pressure will become too numerous and acquire the status of pest for agricultural crops. Thus most important is the balance of nature of which wildlife in an integral and important part.

Abuse and Depletion of Wildlife

One often comes across the headlines in newspapers like 'Last of the blus whales', save the golden langur Tragedy strikes sea turtles, 'major downfall' in rhino population. Great India bustard on the road of extinction...... and so on which narrate the sad story of abuse and depletion of our wildlife heritage. According to an estimate in last 2,000 years the world has lost, by way of extinction about 160 species of mammals and 88 species of birds.

India is endowed with rich biological heritage and her various seripuries and epies are full of vivid and beautiful account on wildlife.. Over 500 mammalian and 1,200 avian species contribute to the riches of the wealth. Unfortunately today the Indian wildlife as that in many other parts of the world, is threatened because of man's unprecedented intervention with nature. Nearly 134 plant species have been declared threatened as against 71,47 and 15 species of mammal, birds, and reptiles respectively (Khoshoo 1984). Great India bustard Choriotis nigriceps confined to the semi-arid regions of Rajasthan and Gujarat, and the white-winded wood dack cairina scutalata inhabiting the Brahamaputra swamps and the flood plains of Assam are both on the birck of extinction.

Wildlife management

The conservation, is the practice embracing preservation, maintenance, substainable utilization, restoration, and enhancement of the natural environment. Wildlife conservation in this sense is the management of nature's living resources if we consider wildlife as consisting both of wild flora and fauna. Since one cannot manage living resources while being indifferent to the non-living environment, the concept of wildlife management becomes the concept of management of natural ecosystems. It also overlaps "forestry" in sofar as it is based on multiple landuse concept and implies management of wild plants as well as wild animals. However, under the present dispensation wildlife management implies management of wild animals as distinct from forestry, and it implies maintenance of habitats suitable to the different species.

In America and Europe wild animals, particularly mammals, are treated as game crops, as do timber in forestry and food crops in agriculture, and hunting is developed as a viable economic activity. In America, wildlife and wildlife related recreation forms multimillion dollar industry. However, in India it is difficult of develop game as in industry because: (i) our forest areas are already overburdened with grazing pressure of domesticated livestock and because of already scarce pasture lands there is little hope of reducing grazing pressure in the forest areas to develop game: and (ii) hunting as a source of recreation is beyond the means of the majority of population.

During past decades the wildlife management implied the understanding, increasing, and controlling of some species. Sanctuaries and National parks were mostly meant of protect a particular species or a group of species and to limited extent for tourism and recreation. Today, however, the emphasis is shifting from the management of species to the management of ecosystem. The objective of wildlife management, as outlined in the World Conservation Strategy, are thus as given below:

i. To maintain essential ecological processes and life support systems in natural ecosystems, on which human survival and development depends. These essential ecological processes and life support systems can be defined to include maintenance of both the biotic and abiotic components of the ecosystems.

ii. To preserve genetic diversity (i.e., the range and variety of genetic material found in the world's wild plants and animals) on which depend the breeding programmes necessary for the protection and improvement of cultivated plants and domesticated animals, as well as much scientific advance, technical innovation, and the security of the many industries that use living resources.

iii. To ensure the sustainable utilization of species and ecosystems. This will include the proper utilization of wild animals for game, forest and grazing, and to achieve proper integration for economy both of rural communities as well as trade and industry.

Fortunately, people have been alerted now and the strong measures being taken in favour of restoring, maintaining, and increasing wildlife population given hope that our rich legacy of wild life will not only be saved for future but developed further. Ways and means employed in wildllife conservation include: (i) protection of breeding stock by enacting game laws and developing wildlife refuges; (ii) artificial stocking including introduction of exotics; (iii) control of predators; (iv) habitat improvement, and (v) game farming.

Minerals and Their Characteristics

Minerals are the nonliving substances obtained from the earth through the process of mining. Formation of minerals took place predominantly in weak zones like mountains, folded and faulted regions, lakes, troughs, and continental shelves through geological processes going on since billions of years. They hold some peculiar characteristic, as discussed below, which distinguish them from other natural resources.

(i) Exhaustibility

Minerals are non-renewable or depleting assets and once mined out the deposit get depleted and lost for ever without any chance of replenishment. Refuse or waste, however, can be consumed afterwards through improved benefication technology.

Simply this exhaustible resource cannot be cropped year after year like agricultural products. Therefore, three distinct stages can be identified in the life cycle of a mine: youth-developmental stage; maturity established stage; and old age-declining ore reserves. If the rate of mineral consumption is increasing fast, the knowledge about more reserves and potential substitutability of many minerals is also being improved. As such these predictions may prove outlived.

(ii) Localised occurrence

Minerals are very localised in their occurrence and the total area occupied them is far limited. Most of the mineral deposits are very sporadic in distribution, for instance, Kolar gold fields of India occupy an area of 8 km. by 1.6 km. formations containing coal and petroleum are, in general, bigger as compared to metalliferous deposits. The coal-bearing formation of India, for example, occupy about 4 per cent of the country's total area.

(iii) Unpredictable occurrence

The quantity and quality of a mineral in any region is not easy to predict. Uncertainties prevail with respect to their occurrence, extent of depth, cost of mining (often linked with depth) and the grade.

(iv) Risk involvement

The process of mining involves large number of problems for miners. Abnormal temperature and pressure variations, rock burst, fire and gas hazard, lack of ventilation and drainage, collapse of roof and side walls are the peculiar problems encountered only in mining industry.

(v) Dependence on other factors

Mineral production is largely affected by factors like the changing needs of consumer industries, availability of stock, fluctuation in prices, and reuse of metallic scrap in production process. A technologically poor country possessing richer grades of ores may have to face keen competition in international market with technologically richer countries may, however, possessing poorer grades or ores since in later case the cost of production is considerably low due to improved mining methods.

(vi) International character

Minerals know no political boundaries. Virtually, no country within its own boundaries has adequate production and reserves of all the minerals needed for its industrial development. The unequal distribution of minerals makes the international trade in its inevitable. Some countries are bestowed with monopolistic rights in respect of certain mineral deposits which result in what is know as geopolitics.

Development and conservation of Minerals

Geologists are of general apprehension that the known reserves of minerals may not last for long and most of them will be exhausted within 100 to 200 years. New possibilities of mineral reserves in land and beyond the land (as in the seabed are being explored which may reduce, if not totally remove, the futuristic fears of exhaustion of the resource.

Presently, limenite, rutile, tin ore, diamond, sulphur, coal petroleum, and calcareous and ferruginous sea sands are largely being mined from the sea-coasts and sea beds. In India, ilmenite, rutile and associated zircon, monazite, and sillimanite are obtained from the beach sands of Madras and Kerala coasts. Coral limestone for industrial use like manufacture of calcium carbide, and calcareous sand for cement are being mined from the sea-bed at Tamil Nadu and Dwarka (Gujarat) respectively for quite long time. Indonesia, Thailand, and Burma recover considerable tin ore from the sea- bed, while England recovers coal and Philippines iron from the sea-bed. A number of petroliferous structure occurring off the coast are being mined for petroleum and its products at several places. The ferromanganese nodules or what commonly known as manganese nodules found over the oceanic floors, generally below 4,000 m depth, are perhaps the most spectacular discovery in the field. The most prolific concentrations of such nodules with high metal values have been found in North Pacific between 139 W to 141 W and 4 N. Although precise information about the reserves of manganese nodules is not available but it is estimated that they may run several billion tons and may prove to be potent source of metals like manganese, nickel, copper and cobalt. It is estimated that 1 million tons of manganese nodules from sea-bed provide about 275,000 tons of manganese, 13,000 tons of nickel, 2,5000 tons of cobalt, and 1,000 tons of copper besides other metals like iron, vanadium, molybdenum, silver and lead. Further, it has now been roughly estimated that the metal resources available in molybdenum, silver, and lead. Further, it has now been roughly estimated that the metal resources available in manganese nodules are several times greater than those available in land, such as cobalt 5,000 times, manganese 4,000 times, nickel 1,500 times and copper 150 times.

The conservation of mineral resources does not mean preservation of minerals under ground to be used by future generations but it implies efforts to utilize minerals in best possible way, depending upon the industrial needs and changing technological requirements, so that the life of the deposits is increased without sacrificing out own needs. From the utilization point of view, minerals are placed under two classes: expendable and non-expendable. Expendable minerals have practically no salvage value and ones consumed are lost for ever, e.g., coal, petroleum, gas and most of the non-metallic minerals. Non-expendable minerals are those which have good salvage value and can be used again and again through recycling, except those quantities that are lost due to corrosion and abrasion, e.g., metallic minerals.

Efforts to conserve mineral resources include:

i. Application of efficient methods of mining to take out every possible tonnage lying under ground.

ii. Utilization of unmarketable ores through the innovative methods or ore dressing and metallurgical practices.

iii. Use and reuse of scraps.

iv. Recovering all associated elements as coproduct or by-product.

v. Economic use of minerals.

vi. Substituting the use or rare land costly minerals with those which are abundant and cheaper.

A facility (house, hospital, port) or even in providing a service to society (medical care, public transport). Firewood, fossil fuels (coal, petroleum and its products, natural gas), solar energy, biogas, wind and hydro power, and atomic power are chief energy resources. Of these, firewood and fossil fuel are being used for supply of energy since time immemorial and are known as conventional energy sources. Wind and hydro power, solar energy, biogas, and atomic energy are the new and non-conventional sources of energy. Fossil fuels being limited in supply may be termed as exhaustible sources while others, like solar, hydel, wind and atomic energy, that would be available to man for infinite period are regarded as inexhaustible sources. Firewood and other biofuels (Such as, biogas derived from biological wastes, domestic garbage etc.) are indeed renewable energy sources but exploitation of fire woods has great ecological implications since we are already short of forests and whatever forests available should not be squeezed or thinned. However, use of other biofuels is known to be the best option in the light of two important acts: their steady of even increasing supply at low cost; their use in energy production reduces pollution of environment.

Because of rapid depletion and poor regeneration of the exhaustible sources of energy the society should concentrate on tapping the inexhaustible and renewable sources under strict economic and ecological consideration.

Conventional and exhaustible energy sources

Fire wood: No one can trace back in the history that who are made the first assault that is on which tree. However, it is evident that man has been logging down the trees for various purposes, including to get firewood as an important one, since long. Because, in olden days man's knowledge and access for various natural resources were limited, he ruthlessly exploited his own surroundings to meet out his day-to-day requirements. The story still continues in various parts of the world at slower or faster pace, but often not because of the dearth of knowledge but out of vested interests or because of having no other option. Felling of trees for firewood is still a common practice in interior where it is the cheapest and, at times the only available source of energy. This had led to thinning or denudation of woodland that needs serious consideration from both ecological and economic viewpoints. Obviously, the only wayout to the problem is making available to the villager the cheaper nonconventional energy sources, such as biogas and solar energy in usable form.

Fossilfuel: Forssilfuels (coal, petroleum, natural gas) are the most extensively used sources of energy today. Like for many other natural resources increase in population and per capital energy demand coupled with the industrialization at a faster pace, are the factors responsible for depletion of fossil fuel world over. According to a report the earth's original stocks of coal and petroleum were about 6,000 billion tons and 1,250 billion barrels respectively, of which by the year 1960,100 billion tons of coal was mined and 100 billion barrels petroleum extracted. The story of natural gas is no different which has acquired the status of irreplaceable commodity of domestic use in modern times. Petroleum and natural gas today fulfil 60 per cent of the world's total energy requirements. It is evident that if the exploitation and consumption of these resources continues at the present incredible rate, there supply may last only for a few more decades.

Non-conventional and inexhaustible energy sources

The rapid depletion conventional energy sources has prompted governments and people to concentrate on finding and tapping some non-conventional energy sources that may last for long. In fact, inexhaustible energy sources, like solar, hydel, tidal, wind and atomic power, can only bring hope for the sustained socio-economic development of humanity for years to come when exhaustible sources will become either inaccessible or too costly to afford.

Solar energy: Research and application in the field of solar energy have opened new vistas in the direction of fulfilling world's future energy requirements. It has especially drawn attention for its practically free steady supply and pollution free use. This energy may be transformed into heat using reflectors and condensers, or into electric energy using solar photovoltaics. The resource can especially be tapped more effectively in the regions where there are long sun-shine hours.

Solar heaters and cookers are becoming popular these days in industry and household. Passive solar architectures provide comfortable housing in extreme cold and hot climates. This engineering has also been successfully applied in warehouse and cold storages. A multipurpose solar dryer-cum-warehouse is under construction at Ganaur (Haryana, India) where chillies, potato-chips and other foodstuffs could be dried and stored. Solar energy has also been effectively used in desalination of water. College of Engineering Pune has designed a solar furnace that can generate heat upto 2,000^{0}Cu which can be send in various metallurgical and other purposes.

Department of Non-conventional Energy Sources of Govt. of India has prepared a perspective plan that envisages generation of energy through non-conventional sources, chiefly solar to the extent of 250 million tons of coal replacement per annum including 15,000 MW of electricity by the year 2001. Indian's first solar power station, soon to be operated near New Delhi, would be a great breakthrough in this direction. Rural electrification through solar photovoltaics is already in operation at some places in the country. Moreover, photovoltaic irrigation pump sets have also been installed at some places.

Wind power: The power of wind since olden days is beings used in running mills, irrigation of fields, and carrying out other farm activities, but its real potential is not so far being realized. According to inventories of Department of Non-conventional Energy Source of Government of India 20,000 MW electricity can be generated from the wind alone in the country. Wind power plants are set up in various parts of the country but they still account for the total capacity of 3.3 MW only. Obviously, this potent resource is still to be tapped optimally.

Hydro power: Use of gravitational flow of water in generation of electricity, what commonly known as hydro power or hydel power is quite common in various parts of the world. India's, first hydro power station, having a potential of 300 KW, was constructed in 1897 at Darjeeling. Later on, many stations with far high potentiality were set up. The country's economically exploitable hydro power resources have been estimated at 41 million KW (or 41,000 MW) of which only 18 per cent of the potential was utilized by 1979. The Brahmaputra basin has the highest hydro power potential and nearly 30 per cent of he country's total; next to it rank the Indus, Godavari, and Ganga basins respectively.

In world, a total $4{,}910 \times 10^9$KW electric power is generated of which hydropower forms of fourth. In highly industrialized countries like USA, USSR, Canada, Japan, Furnace, Sweden, and Norway hydro power is nearly fully developed whereas in the rest of the world it is yet to be developed fully.

Hydel power is considered to be the cheapest source of electricity with the cost of generation ranging from 2 to 3 paise per unit, compared to a minimum of 5 paise per unit of electricity generated by coal based thermal power plants. Although the initial cost of hydel project is generally 25 per cent higher but owing to lower recurring cost, the cost of generation of hydropower is the lowest. A part from such economic considerations, development of such projects is beneficial for irrigation and other purposes too.

Tidal power of water in costal areas of India, however not so far being realized. According to an estimate a tidal power potential of 9,000 MW has been identified of which 7,000 to 8,000 MW is in the Gulf of Cambay, 800 to 1,000 MW in the Gulf of Kutch and rest in sunderbans.

Atomic power (Nuclear power): As already delineated, the fossil fuels beings exhaustible, such sources of energy are not to last for long with the current trend of consumption. Atomic power appears to be the only hope for large scale energy requirements when fossil fuels are exhausted. The reserves of atomic fuels are said to be over 10 times greater than fossil fuels and the advantage of former is that their very small amounts would produce enormous amount of energy. A ton of Uranium ^{235} for instance, would provide as much energy as 3 million tons of coal or 12 million barrels of oil, Atomic energy has its application not only in generation of electricity but has successfully been used in marine vessels, spaceships and even in chemical and food processing industry.

Important constraints in atomic energy generation are: (i) cost of construction and maintenance of plants; and (ii) disposal of radioactive wastes cost of construction of an atomic power plant and its maintenance is very high as compared to any other electric power plant. For running one atomic power plant unit of 235 MW capacity, heavy water of a worth Rs. 10 core is required per year. Further, as a consequence of frequent failures, the running cost of such plants come to be very high.

Moreover, after the working capacity of an atomic power plant is over (in about 25 years), these have to be destroyed so as to avoid radiation spillage. Surprisingly the cost and time for destruction of these plants is almost equal to that for construction. Disposal of radioactive wastes is such a problem that still has no satisfactory solution for long term.

Bioenergy (Energy from biological matter): Organic wastes provide an important renewable source of energy. Biogas generated from the decomposition of organic wastes, like manure and domestic wastes is becoming increasingly popular as a source of heat and light in many parts. It is considered to be advantageous in view of its relatively cheaper supply, and use of organic wastes in its generation reduces the otherwise impending threat of pollution due to their release in environment. As such, it serves two purposes; fuel production and waste disposal. It has more of practical applicability in villages where organic waste in the form of cattle dung, is available in plenty. Biogas so generated is thus economic and convenient to use as compared to conventional practice of burning of dry dung-cakes. Further, the dung-cakes serve only as a source of heat for cooking and allied and purposes while biogas has multifarious applications. The slurry so produces in a biogas plant can well be used as manure for agricultural fields.

2 Environmental Protection

Environmental problems have attracted the attention of a wide cross-section of people all over the world during the last two decades. People are becoming increasingly conscious of a variety of problems like global warming, ozone layer depletion, acid rain, famines, droughts, floods scarcity of fuel, firewood and fodder, pollution of air and water and problems from hazardous chemicals and radiation which have adverse effects on the environment. No nation in the world has been spared nor any citizen untouched. India, at the time of independence, was among the poorest countries with little infrastructure for development. The founding fathers of the nation led by Mahatma Gandhi and Jawaharlal Nehru realised that political independence will have meaning to its citizens only if it helps them to quickly get out of the morass of poverty.

Thus has started, in the words of Indira Gandhi, "an enterprise unparalleled in human history - the provision of basic needs to one-sixth of mankind within the span of one or two generators". The results, in making the infrastructure of the country stronger and in reversing the trend of poverty and its symptoms, have been commensurate with the efforts. The literacy rate has increased to 42.49 in 1991 per cent. Death rate has come down from 4.9 per thousand in 1921 to 10.2 per thousand in 1989. Food production has jumped up by 300 per cent from 50 million tonnes at the time of independence. Every indicator of quality of life has shown betterment.

A disheartening trend

It was in the early seventies that along with the rest of the countries in the world India realised another disheartening trend. The same efforts which helped bring people above the poverty line also put greater pressure on the natural resources of the country.

Irrigation facilities provided also produced salinity in the land. Industrial development provided products but also polluted water and air. Medical facilities improved the health of the people but increased the population, thereby increasing the demand for the products of nature.

Poverty is the fundamental cause which makes people over exploit the natural resources of the country like land, forests and water for meeting their basic needs, for employment, for shelter, for fuel and fodder for their cattle. As Indira Gandhi put it in her address to the Stockholm conference, "Poverty and need are indeed the greatest polluters".

The birth rate in India has come down from 4.1 per cent in 1961 to 3.05 per cent in 1989, but simultaneously the death rate has fallen steeply from 2.74 per cent in the in 1961 to 1.02 per cent in 1989. The present projection is that we will have a population of a billion by the year 2000 A.D.

It is widely recognised that population growth is essentially the result of poverty and lack of education, particularly women's education. Official attitude towards population growth has only a marginal effect. India has to consider population growth as a certainty and make provision for providing the basic minimum needs to an ever-increasing population.

Land degradation

Land is the basic resource for providing the needs of the people. Land degradation is an important environmental problem. It is estimated that nearly 100 million hectares of land, almost one-third of the total area, have been affected by land degradation caused by soil erosion, salinity, alkalinity, wind erosion etc.

Out of the 300 million hectares of land available in the country only about 150 million hectares is cropped. The productivity of the land has to increase from the present 150 MT to 275 MT in order to feed the population by the year 2000.

Soil erosion is a matter of serious concern to India. Deforestation and cultivation of marginal lands have, it is estimated, led to a loss of 6000 million tonnes of soil annually resulting in loss of productivity. The livestock population of India has increased from about 280 million in 1951 to about 375 million.

About 13 million hectares of land is classified as pastures but much of this has been diverted to agriculture by the poor and most of the rest is degraded. Animals have to scrounge for whatever they can get on the fallow lands and on forest lands. Overgrazing has led to further land degradation.

Deforestation

The forests of India occupy an area of about 75 million hectares, i.e. 22 per cent of the land area. These forests supply the demands of people for timber, fuelwood and minor forest products. However, with the increase in human and cattle population the biotic pressures have become intense.

In addition, industrialisation has brought about increase in demand for wood. In the early years of independence the degraded forest land was diverted for agricultural purposes. In addition, development projects such as hydro-electric projects and industries claimed their share of forest land.

A combined result of these causes has reduced the forest cover to about 19 per cent of the area between 1972 and 1982.

The average annual water resource of India is estimated to about 400 million hectare metres (MHM). Out of this, however, about 300 MHM is received by rainfall during the four months of monsoon. The vegetation cover which existed in earlier times used to hold back the monsoon waters and release them steadily after augmenting the ground water.

With deforestation, this cushion is gradually disappearing and the water availability is becoming uneven. A high rainfall would result in a flood while a deviation below the normal would create a drought. Floods carry the soil with them leading to silting of rivers and reservoirs.

This alternating cycle of floods and droughts has been steadily increasing. The annual expenditure on meeting the floods and droughts has reached a level of Rs. 1500 crores.

Industrialisation has posed a threat to water quality through effluents discharged from industries. Untreated human wastes from towns and cities reach the water courses. These discharges affect the quality of surface water and ground water.

The problems of air pollution are beginning to be felt in India with the accelerated industrial tempo. A survey of the cities has revealed problems of excessive sulphur dioxide in some locations. Automobile pollution is becoming a problem in metropolitan cities, particularly in Delhi and Bombay.

India is very rich in biological diversity with about 45,000 species of plants and 75,000 animal species. These genetic resources are the most precious assets which are now threatened by human encroachment. Protection of this natural heritage is an environmental challenge facing the country.

Energy is critical to all development activities covering the household, agriculture, transport and industrial sectors. The rate of commercial energy consumption has more than doubled during the last 40 years, especially in industry and transport sectors.

However, most of the people of India still depend on biomass-based energy such as fuelwood and agricultural wastes. The crisis due to deforestation and indiscriminate cutting down of trees has affected this source of cheap energy.

India's growing population and rapid urbanisation have resulted in an enormous burden on human settlements. By the year 2000, the urban population will touch the figure of 400 million. Planning for this urban colossus is a herculean task. The shortages in housing, potable water and sanitation are already enormous and will grow further unless remedial action is taken in time.

It does not, however, mean that all developmental activities must be halted or that we should go back to a simplistic life style. The provision of minimum needs of food, shelter, education and health to our vast masses demands that the basic industries and infrastructures such as agriculture, power, transport, communications, steel, fertilisers, pesticides, cement etc. be established.

One sees no option but to adopt the modern concepts of development, of course, after ensuring that these developmental patterns do not damage ecological balance as damage would not only hamper further development but also result in greater poverty for the poor who depend on the environment for food fuel, fodder and shelter.

For the sustenance of a socio-political system aimed at people's welfare, protection of environment is as important as economic development. The two concepts are totally inter-dependent and form the essence of the concept of 'sustainable development'. Our country's ingenuity and creativity should, therefore, be directed towards identifying a developmental path which is harmony with environmental considerations.

The responsibility to harmonise environmental protection with development objectives rests with the Ministry of Environment and Forests. It serves as a focal point for the planning, promotion and coordination of various environmental and forestry programmes.

The task is sought to be fulfilled through action on diverse fronts such as survey and conservation of natural resources, impact assessment of development projects, incorporation of environmental protection in the planning process, control of pollution, regeneration programmes, legislation, research, extension, education, training, dissemination of information and creation of environmental awareness.

The State Governments, Central and State Pollution Control Boards, the Indian Council of Forestry Research & Education, the Wildlife Institute of India, the Botanical and Zoological Survey of India and a host of other agencies are partners in the activities related to environment protection. A number of initiatives have been taken in the past few years to put environmental protection on firm ground.

Forest Conservation

Forests play an important role in maintaining

environmental stability and in supplying the essential requirements of the people on a renewable basis. Over the years the forests have suffered depletion due to relentless pressures arising from the increasing demand for fuelwood, fodder and timber; inadequacy of protection measures; diversion of forestlands to non-forest uses, and the tendency to look upon forests as a revenue earning resource.

Realisation regarding importance of the forests has always been with us, but the first practical step came in the shape of the Forest Conservation Act, 1980 which prohibited diversion of forest lands for non-forest purposes, except with the previous permission of the Central Government.

Under this Act, such permissions are generally refused. Where unavoidable for development purposes, permission is given only under condition of compensatory reforestation. This Act has been amended in 1988 to make the provisions more stringent and to enforce penalties even against officers who facilitate violation of this law.

As a result of strict enforcement, the rate of diversion of forest land has been brought down to about 16,000 hectares per annum compared to the rate of 1.5 lakh hectares, per annum prior to this Act.

Other efforts to conserve forests include promotion of improved energy efficient stoves and alternative energy source. Modern forest fire control methods have already been adopted by some States to control forest fires.

In 1985 the National Wastelands Development Board was set up to promote afforestation. The objective was to bring five million hectares of wasteland every year under fuelwood and fodder plantations with the participation of the people.

The Board has succeeded in afforesting 7.16 million hectares till the end of March 1989 through its programmes on social and agro forestry. While this achievement is substantial, the target of five million hectares per year is yet to be attained.

In order to achieve the same, a comprehensive technology mission will be launched shortly. This technology mission will marshal the scientific and technological inputs available in our country to reclaim saline and alkaline lands and other kinds of wastelands to afforest the upper catchment areas of our important rivers and to meet the fuelwood and fodder requirements of the people.

Wildlife Conservation

The Wildlife (Protection) Act, 1972 has been amended to prohibit trade in endangered species and derivatives thereof. The National Wild life Action Plan has been implemented for a rational and modern wildlife management. A comprehensive report, "Planning and Wildlife Protected Area Network in India", has been prepared and released.

The scheme "Project Tiger" has been expanded for the maintenance of a viable population of the tigers in India, with 17 Tiger reserves in 13 States of the country covering an area of 26,643 sq.km.

National forest policy

The Natioanl Forest Policy adopted in 1952 aimed at forest coverage of one-third of the total land area of the country. But this could not be attained for various reasons and the process of deforestation continued. This has necessitated area-examination of the forest policy. The revised National Forest Policy was finalised after extensive consultation with the States, experts and all relevant agencies.

The salient feature of the new forest policy are: maintenance of environmental stability through preservation and restoration of ecological balance; conservation of the natural heritage of the country by preserving the remaining natural forests and protecting the vast genetic resources for the benefit of posterity; meeting the basic needs of the people, especially fuelwood, fodder and small timber for the rural and the tribal people and maintaining the intrinsic relationship between forests and the tribal and other poor people living in and around forests by protecting their customary rights and concessions on the forests.

Pollution control

India faces serious problems of environmental pollution. Water pollution from untreated community and industrial wastes flowing into our rivers and streams flowing into our rivers and streams, increasing use of fertilisers and pesticides, dumping of organic and inorganic wastes etc. are issues that need to be tackled on priority. Air pollution, long regarded as the bane of industrialised nations looms large over our country.

The main instruments for control of pollution of water, air and noise are the Water (Prevention and Control of Pollution) Act, 1974, the Air (Prevention and Control of Pollution) Act, 1981, the Water Cess Act, 1977 and the Environment (Protection)Act, 1986.

The Water and Air Acts have been amended in the recent years to make provisions more stringent. The Pollution Control Boards have also taken up systematic monitoring of air and water quality through a network of quality monitoring stations-3.6 for water and 14 for air.

Out of the nearly 4,000 major and medium polluting industries, more than half have already installed pollution control equipments. Standards have been notified for 26 priority industries. So far 3,300 prosecutions have been launched by the Pollution Control Boards against recalcitrant industries.

Ganga Action Plan

More than two-thirds of the pollution in our rivers is caused by human and cattle wastes. In 1985 the Government took up a masssive project to clean up Ganga, the most important river of India.

A Central Ganga Authority has been set up and an action plan was drawn up for cleaning the polluted stretches of the river Ganga by intercepting, diverting and treating about 900 mld. of the domestic waste water generated along the river to conform to effluent standards.

Schemes of low cost sanitaton, river front development, construction of electric crematoria etc. have keen taken up. Two hundred and sixty-two schemes have so far been sanctioned under the Ganga Action Plan at a cost of Rs.258.44 crore covering the States of Uttar Pradesh, Bihar and West Bengal. Fifty schemes have already been completed. Extensive awareness and public involvement activities have been taken up involving youth, students and pilgrims.

The Indian tradition of conservation can be traced back to pre-historic times. The theme of conservation prevades Indian culture, art and all its religions, Hindu, Muslim, Christian and Sikh. From Indira Gandhi's address to the Stockholm conference in 1972 to Rajiv Gandhi's address to the United Nations in 1988, and Environment Minister Shri Kamal Nath's address in Rio conference in 1992 India's leaders articulated the environmental concern of India effectively.

In India the last decade had seen the rise of public

involvement. Silent Valley, the Doon Valley limestone quarries, the Chipko Movement, the protests against Narmada and Tehri...these mark a beginning.

In response to them came a few legislative measures-the Environment Protection Act, the strengthening of the Air and Water Acts, the new Forest Policy...all these followed the creation of an environment ministry which started off as a lame duck waiting-room ministry for politicians and now, is being taken seriously. As the environmental problems have already reached an alarming situation. The concern for environment is felt by people all over the globe. Environmental Impact has become the focus of development programmes.

Response to Environmental Challenges

Environmental problems in India can be classified into two broad categories:

(a) Those arising from conditions of poverty and under development.

(b) Those arising as negative effects of the very process of development.

The first category has to do with the impact on the health and integrity of our natural resources (land, soil, water, forests, wild life, etc.) as a result of poverty and the inadequate availability, for a large section of our population, of the means to fulfil basic human needs (food, fuel, shelter, employment, etc). The second category has to do with the unintended side-effects of efforts to achieve rapid economic growth and development. In this latter category would fall distortions imposed on national resources from poorly planned development projects and programmes, as from lack of attention to long-term concern by commercial and vested interests. Thus, it is clear that a concern for environment is essentially a desire to see that national development proceeds along rational sustainable lines. Environmental conservation is in fact the vary basis of all development.

In India the over-riding concern for environment protection has been stressed in the Directive Principles of the Constitution of India. The relevant provisions are as follows:

The State shall endeavour to protect and improve the environment and to safeguard the forests and wild life of the country (Art.48A).

It shall be the duty of every citizen of India:

(g) to protect and improve the natural environment including forests, lakes, rivers and wild-life and to have compassion for living creatures; (Art. 51A)

There are more than 200 pieces of legislation in India that directly or indirectly relate to the protection of environmental resources. Concern for environmental protection is also stressed in three policies of economic development. Successive plan documents have given recognition to this important aspect.

To integrate the environmental concerns in the planning process, a high level National Committee on Environmental Planning and Co-ordination (NCEPC) was set up in 1972. The Committee had done valuable work in a number of areas related to environmental planning. These include environmental appraisal of development projects, human settlements planning, formulation of environmental guidelines and creating environmental awareness at various levels.

In the recognition of the need for a fresh comprehensive look at the administrative and legislative aspect of environmental protection, the Government of India constituted a High Power Committee under the Chairmanaship of the Deputy Chairman of the Planning

Commission, Shri. N.D. Tiwari. The Tiwari Committee as this Committee is popularly known submitted its report to the Prime Minister in September, 1980.

National Environmental Advisory Committee

A National Environmental Advisory Committee was constituted in 1983 for highlighting environmental issues and giving advice on remedial action; raising public discussion on national issues of environmental significance, promoting public commitment and participation in environmental programmes, and providing feedback on people's perception of national environmental priorities and action plans.

Report of Tiwari Committee

The Tiwari Committee had identified the following major areas of environmental concern;

(a) *Environmental Pollution*: All our rivers, lakes and ponds are heavily polluted by the discharge into them of untreated sewage and industrial effluents. The pollution constitutes a most dangerous health hazard apart from creating adverse effects on fish production and recreational possibilities, etc. It is also known that only a microscopic proportion of our industries are at present adequately treating their liquid effluents before discharging them into natural outfalls.

The pollution of air by untreated gaseous effluents from industrial plants and by the exhausts of transport vehicles also constitutes a serious health hazard in many of our bigger towns.

Noise pollution has became a problem in many of our bigger towns.

(b) *Land and Water Resources:* The mismanagement of our land and therefore also our water resources is of

> awesome scale. According to estimates made by Ministry of Agriculture in March 1980 as much as 175 million hectares (mh) out of the country's total land area of 304 mh, for which records exist, are subject to environmental problems.

In a study made in 1972, it was estimated that on an average, India was losing about 6000 million tonnes of top soil per annum through water erosion and that these represented in terms of major nutrients NPK alone, an annual loss of Rs. 700 crores. The corresponding loss today must be of a much higher order.

Again, according to the Report of the National Commission on Floods (1980) the losses on account of floods average to over Rs. 100 crores per year. The total area subject to periodic floods which was estimated at 20 million hectares in 1971 now stands at the level of 40 million hectares.

Soil erosion also causes the premature siltation of tanks and reservoirs. The threat s of particular seriousness because in most cases of multi-purpose projects alternative sites for storage are just not available even if we can find the large sums of money needed to build new reservoirs in place of those which go out of commission.

The extent of forest cover is a good indicator of the health of the land. The large scale deforestation in recent decades had rendered the sensitive catchment areas in the Himalayan and other hilly areas particularly vulnerable to soil erosion. Poor land management thus aggravated the problems of droughts and floods. No more than about 12 per cent of the country's land surface is actually under adequate tree cover as against the target of 33 per cent prescribed by the National Forest Policy of 1952. Further, although 13 million hectares are classed as

"Permanent Pastures", these areas are in fact generally without any vegetation on account of either overgrazing or encroachments.

Mention must also be made of the damage caused to our agricultural land in canal irrigated areas by water-logging and consequent salinisation on account of our failure to provide them with adequate drainage. It is estimated that about 6 million hectares of goodlands are affected by varying degrees of salinity.

The problems are so formidable in nature and scope that the Committee recommended "as a first step the setting up of Capital Land Commission" to look into issues concerning the health and scientific management of our total land resources.

(c) *Natural Living Resources:* With the inadequate level of scientifc knowledge in the country, it is not known how many species of flora and fauna are threatened, endangered or extinct; but it must be considerable considering the very rapid shrinkage of all natural forests and other ecosystems throughout the sub-continent in recent years.....The diversity of biological organisms a vital resource which needs to be carefully protected in natural ecosystems if we are not to close many possible evolutionary options for benefitting future generations. There is inadequate knowledge and understanding of our country's valuable marine ecosystems and total absence of protective measures has led to the wanton exploitation and destruction of these resources.

(d) *Human Settlements:* It was not possible for the Committee, within the limited time available, to go into these aspects in any details. Therefore, it suggested that a separate group be constituted to study this subject area and its implifications for the environment.

However, the major problems of human settlements in India are well-known. Leaving aside the elitist residential and commercial areas in some of our bigger towns, the general condition of our human settlement is, to put it mildly, far from satisfactory from almost every point of view. While it would be futile to expect any significant improvement in these conditions for many years, and perhaps decades to come, there are two matters which we cannot afford to neglect any longer. These concern the indescribably poor sanitary conditions which prevail in almost all our villages and towns and our failure to supply clean drinking water to the vast majority of our people. Filth and dirt, and open drains and stinking sewage, along with the swarms of flies and mosquitoes which they breed, are to be seen everywhere. The supply of safe drinking water has become a problem even in our biggest towns, so much so that even cities like Delhi and Bombay are visited by epidemics of jaundice and other water-borne diseases every now and then.

(e) *Environmental Awareness and Education:* A sound, well-informed interest in nature among our people needs to be cultivated. Nature education, at all levels of education and public communication, must be efficiently organised especially in our schools.

New Legislative requirements

Some of the important legislative measures suggested by the Committee for recommending Legislative Measures and Administrative Machinery for ensuring environmental protection are as follows.

(i) Biosphere reserves: The Government may consider comprehensive legislation to give effect to the objectives to be served by Biosphere Reserves. Such legislation should include provisions for designating areas characterised by considerable genetic diversity in economic plants or are valuable for serving as "hot spot screening nurseries" for screening material for the detection of genes for resistance to pests as "Genesanctuaries".

(ii) Protection of Grazing Lands: The States be persuaded to enact suitable legislative support for the protection of Grazing Lands.

(iii) Protection of Endangered species: Supporting Indian legislation to effectively implement the Convention on International Trade in Endangered Species of Wild Fauna and Flora (CITES) to which India is a party.

(iv) Toxic Substances Control Act: This is required to ensure the protection of the environment from the sale manufacture, use and disposal of chemicals other than pesticides.

(v) Scientific Land Use: The existing regulations relating to the use of fertile soil for brick-making mining etc. require careful scrutiny. Suitable legislative provisions will have to be made to prevent the growing abuse of land and for enforcing strict reclamation regimes after the soil is removed for brick-making.

(vi) Prevention of noise pollution: Comprehensive legislation is required to cover the outdated bits and pieces of related unimplemented legislation at present on the statute books.

(vii) Prevention of denudation of forests: Legislative measures should be devised to curb the degradation caused by the greed of commercial interests through severe punishment while appropriate steps should be taken to meet the needs of those below the poverty line for fuel and fodder.

The Administrative Measures Suggested by Tiwari Committee

(i) The committee expressed the need for creating a Department of Environment (DOE) at the Centre under the charge of the Prime Minister to provide explicit recognition to the pivotal role that environmental conservation must play for sustainable national development. (The recommendation was accepted and a Department of the Environment duly constituted in 1980. The Department of Environment is the focal point within the Government of India for environmental management. The Department of Environment has, in addition to its catalytic promotional role in environmental management at the national level, executive functions in the field of pollution control. It also undertakes major programmes for ecological restoration.)

(ii) In order to conserve the nation's ecological base it was recommended that a Central Land Commission should be set up. (The recommendation has since been accepted by the Government and steps are being taken to implement it.)

(iii) Action taken to preserve/improve the environment.

The steps to preserve/improve the environment have been as follows:

(a) *Pollution Control:* The Central Board for Prevention and Control of Water Pollution (which is under the administrative control of the Department of Environment), together with the State Pollution Control Boards, completed a country-wide rapid inventory of pollution from large and medium industries. This was in execution of the "Control of Pollution at Source" programme. Minimal National Standards (MINAS) for polluting discharges from specific industries were formulated, and control measures implemented in a progressively stringent manner. About thirty percent of large and medium industries of the country have installed pollution control equipment. A network of about 120 monitoring stations to check water pollution has been established. Zoning and classification of all the 14 major inter-state rivers have been completed to provide a basis for water quality management. The river basin-wise inventory for the Yamuna and Ganga has also been completed.

The Air Pollution Control Act became effective from May, 1981. Programmes in progress, after the framing and notification of rules under the Act completed, include strengthening of administrative support system, inventory of area to be declared air pollution control zones, evolution of ambient air quality standards as well as industry specific emission standard.

Projects in fields such as industry, energy, irrigation and mining have been subjected to environmental impact assessments, on the basis of information provided by the project authorites and supplemented by the Department of Environment's information gathering machinery.

(b) *Land Resource Management:* Notable efforts have been made to plan, organise and provide funds to improve desert environments. To improve ecology as also to meet the requirements of the local cattle and human population forestry has been taken up relatively extensively. Rural electrification with a view to assisting the exploitation of ground water has been promoted. Construction of Rajasthan Canal envisages to transform barrren un-inhabited areas of western Rajasthan into a vast granary.

With a view to meeting the basic and economic needs of the community, the programme of Social

Forestry-which inter alia includes Farm Forestry. Extension Forestry Reforestation in degraded forests and Recreation Forestry-to augment the supply of fuel wood, fodder and small timber to rural areas was taken up in the Sixth Plan to cover 100 chronically deficit districts.

The new 20-points Economic Programme of the late Prime Minister included afforestation, social forestry and farm forestry.

A National Eco-Development Board has been set up to achieve sustainable eeconomic and social development through national utilization of natural resources-water, soil, plant and animal life.

In order to demonstrate methodologies for reclamation of environmentally degraded areas, to create general environmental awareness and to involve public participation in the solution of environmental problems a number of projects have been carried out by the Department of Environment under its Eco-Development Programme.

Universities in several States and voluntary agencies organised Eco-Development Camps. Task Forces of ex-service personnel for eco-developmental work in Shivaliks (UP), Rajasthan Canal etc. have been raised. Programmes have also been formulated for extension of eco-developmental work for mined area reclamation, environment in metropolitan settlements, hill resorts, pilgrimage centres and the Himalayan foothills, etc.

(c) *Natural Living Resources:* India has initiated the formulation of a National Conservation Strategy. It has also adopted a National Wild Life Action plan during the year 1983-84, 13 sites in different biographical regions have been identififed for designation as Biosphere Reserves. Work on environmental impact assessment, wildlife conservation and faunal composition of different eco-systems has been taken up. Studies on faunal composition of marine, estuarine, fresh water and terrestrial eco-systems have been conducted in some areas. Activities to survey, identify and investigate plant species are underprogress. The Botanical Survey of India (BSI) has set up a countrywide network with regional offices. The herbaria of the survey hold over 2 million plant specimens. More than 150 new plants have been discovered.

Survey of unexplored areas, intensification of the District Flora Programme, preparation of comprehensive documents on new flora of India and study of endemic and treated plants were among the major programmes. It has also been involved in assessment of environmental impact of several developmental projects and inventorisation of plant resources in important eco-systems. These include the Silent Valley, Sutlej-Beas Link, Tehri Dam, Lalpur Dam, Idukki Valley etc. and National Parks. Games sanctuaries and areas to be designated as Biosphere Reserves.

The Zoological Survey of India has primarily been engaged in the survey of faunal resources and collection of base-line scientific data with regard to their taxonomy bio-ecology animal behaviour and pollution, etc. Special emphasis was given to the studies on the ecology and population of rare and endangered animals and birds. Studies on final composition of marine, estuarine, fresh water and terrestrial eco-systems were conducted.

(d) *Environmental Awareness:* Programmes have been addressed to specific target groups such as planners and decision-makers, administrators, executives, students, etc. For higher level administrators, senior executives and planners a series of carefully structured workshops were organised at the

Administrative Staff College of India, Hyderabad, the Indian Institute of Management, Bangalore and the Indian Institute of Public Administration, New Delhi. To generate environmental awareness in the public, nation-wide celebrations of World Environment Day and Wildlife Week are organised every year. The first National Environmental Congress and first National Conference of Legislators on Environment have also been held. The National Museum of National History (NMNH) further expanded its facilities for non-formal environmental education, and catered to wide spectrum of the public with particular emphasis on children, and the disadvantaged and handicapped segments of society, in coordination with the States and Union Territories.

(e) *Promotion of Environmental Research:* Promotion of research has been one of the major activities aimed at developing information on various aspects of environmental management and for building up of scientific manpower to deal with the environmental problems. The projects are conducted by a number of universities, R&D institutions, non-governmental agencies and field formations of the Department of Environment. Monitoring of environmental parameters, pollution control, energy use, environmental aspects in human settlements, instrumentation, effects of pesticides and impact human activities on natural eco-systems are among the subject areas dealt with in such projects.

A distinctive feature in the research programme was the introduction of coordinated research projects in selected issues of national importance. These include an All India Coordinated Project on Heavy Metal, a multi-disciplinary project of Microbial Degradation of

Industrial Waters, a coordinated project on Ethnobiology and multi-disciplinary studies for monitoring of environmental effects of the Idukki River Valley Project and Beas Sutlej link. In addition, under the Integrated Eco-Development Programme, action-oriented projects on identified priority areas have also been launched in the Himalayan Region, Western Ghats and the Ganga Basin.

As part of its programme to create a strong research base and training support, a number of centres of excellence are being set up. One such Centre for Ecological Research and Training with particular reference to the problems of the Western Ghats region has been set up at the Indian Institute of Science, Bangalore. An Institute on Himalayan Environment and Development is also in the process of being set up.

(f) *Environmental Impact Assessment and Monitoring:* Environmental appraisal is an important responsibility assigned to the Department of Environment. This involves evaluation of environment implications and incorporation of necessary safeguards for such activities having a bearing on environmental quality. Questionnaires for eliciting information on environmental aspects of development projects in selected sectors, such as industry, multipurpose river valley schemes, thermal power generation, shipping and harbours, and mining have been prepared. The project authorities are required to incorporate a chapter on environmental aspects in their feasibility reports.

(g) *Centres of Excellence:* Strengthening active research training groups in universities/institutions working in areas of interest is necessary. The Department of Environment proposes to selectively support such groups in priority areas of environmental science and management, such as mangrove ecosystem, environmental education, mine environment studies, etc.

The Ecological Research and Training Centre is one such centre already functioning at the Indian Institute of Science, Bangalore, where two scientific programmes have been initiated, viz., one on ecological science focusing on eco-development-oriented research on the Western Ghats and another on environmental studies programme focusing on training with special reference to air and water pollution.

(h) *Environmental Information:* The scheme for setting up of the Environmental Information System (ENVIS) with national network of Distributed Information Centres (DIC) and a focal point in the Department of Environment was approved in the Sixth Plan period. It has become operational through the installation of a Computer Terminal linked with the computer system of the National Informatics Centre (NIC) to serve as the focal point facility for storage, retrieval and dissemination of information to the users. Distributed Information Centres were set up at the Industrial Toxicology Research Centre, Lucknow, the Central Board for Prevention and Control of Water Pollution, New Delhi and 7 other centres for information services in specific subject areas. Centres in respect of other subject areas are being set up.

(i) *Pitambar Pant National Environment Fellowship:* The Pitambar Pant National Environment Fellowship Award was constituted by the Government of India in 1978 to encourage and recognise excellence in any branch or research related to the environmental sciences. The fellowship is awarded every year on the World Environment Day, i.e. on June 5. The fellowship is in recognition of significantly important research/development contri-bution and is also intended to encourage talented individuals to devote themselves whole time to R & D pursuits in the field of environmental sciences for the duration of the fellowship, which is two years.

International Cooperation and Legislation for Environmental Protection

(i) International Cooperation

In the period between the 1972 United Nations Conference on the Human Environment and now, the environment has moved into the mainstream of issues that concern the international community. In the UNEP Government Council, and its numerous other areas, nations have found that they could reach a broad consensus on nature, magnitude and importance of international problems and measures needed to tackle them. Countries have a stake in each other's environmental welfare and stability. There is a massive scope without any impairment of sovereignty and countries legitimate rights for cooperative action on sound management of national resources. Agreement is already present between various groups of governments on a wide range of environmental issues. There is a perception of mutual benefits, UNEP's Regional Seas Agreements, the World Soils Policy, the draft Convention on the Protection of the Ozone Layer, the Trade in Endangered, Species, Wetlands and World Heritage Conventions between and among developed and developing countries are only a few examples. There are 113 conventions and protocols listed in UNEP's register of international treaties and other agreements in the field of the environment.

The principal focus for the dialogue on environment among developing nations has been the meetings of the non-aligned countries and the Group of 77. The UNEP Governing Council itself has provided a major forum for co-operation among the members of the Group of 77. Typical of the Governing Councils the prevailing spirit of cooperation is the provisions of US 1 million dollar by the Government of Argentina-to be used through UNEP's catalytic role-for technical assistance to help developing nations address serious environmental problems. Outside the Governing Council the South Asian Cooperative Environment Programme provides another example of mutual assistance among developing countries.

For the significant part of the last decade, the dialogue among developed countries focused largely on environmental pollution control provides an example. But in recent years there has been a marked shift to adopting preventive policies and taking preventive measures. Notable achievement have included the adoption of the Convention on Long-Range Trans-boundary Air Pollution and OECD's guiding principles concerning international aspects of environmental policies and recommendations and decisions concerning the export and import of banned or severely restricted chemicals.

The decade since Stockholm has also witnessed an increase in environmental cooperation among the nations of the Council for Mutual Economic Assistance (CMEA). This has been accompanied by an increasing emphasis on environmental protection in the cooperation between CMEA Member States and developing countries which has included such topics as hydropower, water and land management and assistance with the installation of environmentally sound new plant and machinery.

Environmental issue may need to be dealt with on different geographical scales, as well as on different time-scales. In other words, they may arise, and/or require action, at the global, regional, sub-regional or other levels. Problems of global concern, including those of the joint management of the global commons and of environmental interdependence, can be dealt with in the overall setting of multilateral internaional negotiations. A case in point is the lay of the sea. There are also clearly identifiable issues characteristic of given geographical settings and countries, which are best dealt with regionally, sub-regionally or nationally, even though many of these have a broader significance.

(ii) National legislation

Since early 1970s there has been a massive expansion in the establishment of national environmental bodies (agencies, departments, councils, committees etc.). There has been considerable increase in the number of governmental and non-governmental bodies dealing with environmental issues. It is estimated that there was about 2,500 non-governmental organisations in 1972, the number rose to about 15,000 in 1981. The period saw also bursting legislation dealing with environmental issues both in developed and developing countries. In OECD nations, for example, whereas the 4 major national environmental laws were passed in 5 years 1956-60, 10 in 1961-65 and 18 in 1966-70, between 1971-1975, 31 such measures were adopted. and a further 25 followed in the 4 years 1976-79 inclusive. By the end of the 1970s statistics of reports on the state of the environment (or both) were being produced in many countries (for example, Germany, Federal Republic of Finland, Japan, Philippines, Spain, Sweden, United States of America).

(iii) Public Participation

The Stockholm Conference provided on impetus renewed interest in environmental education during the 1970s. In response to the Stockholm Recommendations, UNESCO and UNEP set up an International Environmental Education Programme which aims at promoting exchange of information and experiences in the field of environmental education.

Public participation in environmental affairs has increased. Some of the interest was spurred by the efforts of the media to cover the field.

Environmental awareness increased most in wealthier and better educated populations and stimulated by books and films as well as newspapers, magazines and the electronic media/ Non-governmental organisations played an important role in the development of a better understanding of environmental issue. Public organisations affected the decision-making process in a number of countries, particularly in connection with nuclear power development.

On the whole, progress was apparent in both environmental education and public awareness of environmental issues during the decade.

(iv) Position in India

International Cooperation

India's Department of Environment functions as the nodal agency for United National Environment Programme (UNEP), the South Asia Cooperation Environment Programme (SACEP) and the International Union for Conservation of Nature and Natural Resource (IUCN). India has been actively participating in the various sessions of UNEP with a view to ensure that programmes are more relevant to the developing countries.

The South Asia Cooperative Environment Programme has recently set up the inter-governmental organization with Afghanistan, Bangladesh, Butan, India, Maldives, Pakistan, Sri Lanka and Iran as its members. At the 1981 ministerial level meeting, areas were identified for implementation.

India was assigned the focal point responsibilities in

the areas of Environment Education, Environmental Legislation and Wild Life Management. At the 1983 Government Meeting of SACEP number of projects have been identified for implementation in the area of Energy, Environmental Education, Legislation and Environmental Impact Assessment.

India has a number of on-going Bilateral Programmes with both developed and developing countries.

The Constitutional directives (Art.48 and 51-A) and development policy have provided a strong base for enactment of legislative measures as required for environmental protection. There are several laws enacted from time to time which are directly related to environmental protection. Among them more recent ones are the Insecticides Act, 1968, Wildlife Protection Act 1972, Water (Prevention and Control of Pollution)Act 1974, Water Pollution Cess Act 1977, Forest Conservation Act 1980 and the Air (Prevention and Control of Pollution) Act, 1981.

3 Air Pollution

In recent years, great concern has been universally voiced regarding environmental pollution arising as a side effect of industrial activities. In our national situation with the growth of industries, more and more toxic substances are either used as raw materials or given off during manufacturing processes in the form of dust, fumes, vapours and gases. These pollutants ultimately dissipate in the working environment and pose occupational health hazards. According to a U.S. Estimate, there are some half a million man made substances already present as pollutants in our total environment. These pollutants, when get into the air and water, disturb the ecological balance of nature. Control of air contaminants is therefore essential for we are just the custodians of environment for our future generation,. This can be achieved by properly recognising, evaluating and controlling the industrial environment. Considering the above, we in India, may well be satisfied with our consciousness in this matter and it may be mentioned that several Air and Environmental Authorities have been in existence from the 1960-65 period onwards. An Indian 'Water Pollution Bill' came into force from 1969 and a National Committee on Environmental behaviour started functioning from the early seventies. As a result of increasing awareness, four more acts came into existence which are the water (prevention and control of pollution act, 1974, the air (prevention and control of pollution) cess act 1977, the air (prevention and control of pollution act, 1981) and the environment protection act, 1986.

The dry air contains 78.09% nitrogen by volume and 20.94% oxygen the remaining 0.97% is composed of gaseous mixture of carbon dioxide, helium, argon, krypton, nitrous oxide and xenon, as well as very small amounts of some other organic and inorganic gases whose amount in the atmosphere vary with time and place. Various amounts of contaminants continuously enter the atmosphere through both natural and man-made processes that exist upon the earth. That portion of these substances which interacts with the environment to cause toxicity, disease, aesthetic distress, physiological effects or environmental decay, has been labelled by man as a 'pollutant'.

In general, the actions of people are the primary cause of pollution and as the population increases, the attendant pollution problems also increase proportionately. The first significant Change in man's effect on nature came with his discovery of fire. Prehistoric man built a fire in his cave for cooking, heating and to provide light. The problem of air pollution came into existence at this time.

The British Parliament passed an Act in 1273, forbidding the burning of coal in London because it was beginning to choke the atmosphere. In 1300, King Edward I issued a royal proclamation, "Whoseever shall be found guilty of burning coal all suffer the loss of his head." In 1306, a man was executed for violating this regulation. Later, the law fell into disuse as the industrial revolution took place in England.

Today, the natural terrain that surrounds large cities is recognized as having a significant bearing on the problem of air pollution. However, this is not an altogether new concept either. Historians tell us that Los Angeles which in recent years has become a national symbol of comparison for excessive smog levels, was known as the 'valley of smokes' when the Saniards first arrived there.

It has been found that a significantly increasing volume of particulate matter entering the atmosphere, scatters the incoming sunlight. This reduces the amount of heat that reaches the earth and tends to reduce its temperature. This decreasing mean global temperature of recent years have been attributed to the rising concentration of air-borne particles in the atmosphere. A counter acting phenomenon commonly referred to as the "green-house effect" is caused by the increasing amount of carbon dioxide found in the atmosphere. It has been estimated that if carbon dioxide content in the atmosphere generated in combustion processes continues to increase at the present rate, the mean global temperature could rise by 4°C in the next five decades. There has been conjecture that this might become a matter of great importance because small temperature increase could cause a partial melting of the ice caps of the earth causing continental flooding and devastating effects on man.

Air pollution can cause death, impair health, reduce visibility, bring about vast economic losses and contribute to the general deterioration of both our cities and country-side. It can also cause intangible losses to historical monuments such as the Taj Mahal which is believed to be badly affected by air pollution.

On account of large scale industrial activities in advanced countries notably the USA, UK, and other European countries, fall of acid rain has been reported in Scandinavia. This has reduced forest growth in Scandinavia. In Canada, thousands of lakes have been destroyed due to acid rain. Apart from the international issues involved, the basic ecology is affected. Large scale deforestation apart from creating an imbalance in the oxygen proportion of the atmosphere, affects weather and rain patterns as well. Industrial activity, particularly in thermal power stations, cement plants, oil refineries, chemical complexes, metallurgical industries, steel plants and fertilizer complexes, causes major, problems of air pollution. The effects of air pollution are felt more by the elderly and infants as chest and respiratory complaints are very common among them. Tragic instances of death have also been reported in air pollution episodes such as the London smog of 1952, and the Bhopal gas tragedy of 1984.

It is, therefore, a matter of great importance that engineers of all disciplines consciously incorporate in their designs sufficient constraints and safeguards to ensure that they do not contribute to atmospheric pollution, in addition, they must apply their ingenuity and problem-solving abilities to eliminate air pollution where it exists and restoring the natural environment.

There are three methods of identifying air pollution:

1. Sensory recognition
2. Physical measurements of pollution.
3. Effects on plants, animals, and buildings.

These methods are widely used and they have their merits and demerits.

1. Sensory recognition—Usually the first awareness of an air pollution problems is through some effects on the individual. These are:

(a) strong or unusual odours
(b) Reduction in visibility
(c) Eye irritation
(d) Acid taste in the mouth

(e) Feel of grit under foot

There are highly subjective phenomena and vary from individual.

(2) Physical measurement—While sensory perception may provide the first indication of the presence of most of the contaminants in the air, it is often not possible to detect trace quantities of many air-borne toxic substances or the presence of radioactive matter through the senses. Their identification required physical measurement by standard methods of sampling and analysis.

(3) Effects on plants, animals and buildings—Effects of air pollution can be observed on the growth of plants and health of animals. Similarly, its deleterious effect on building can also be observed. Thus plants, animals, and buildings act to some extent as indicators of certain atmospheric impurities.

Classification of pollutants

Air pollutants can be classified as follows:

1. Natural contaminants
   e.g., natural fog, pollen grains, bacteria, and products of volcanic eruption
2. Aerosols (particulates)
   e.g., dust, smoke, mists, fog and fumes.
3. Gases and vapours

The various gases and vapors, which are important air contaminants are given in Table 1.

Table 1: *Air Contaminants*

No.	*Group*	*Examples*
1.	Sulphur compounds	SO_2, SO_2, H_2S, mercaptans
2.	Nitrogen compounds	NO, NO_2, NH_3

3.	Oxygen compounds	O_3, CO, CO_2
4.	Halogen compounds	HF, HCI
5.	Organic compounds	Aldehydes, hydrocarbons
6.	Radioactive compounds	Radioactive gases

In addition to the above, studies by various investigators have indicated, that some of these contaminants undergo chemical reactions when they enter the atmosphere. As a result the end products formed are more harmful than the original contaminants. For example, it is reported that unsaturated hydrocarbons react with nitrogen dioxide in sunlight to form smog.

Natural Contaminants

Among natural contaminants pollen is important because of its peculiar properties irritating to some individuals. Pollen grains are the male gametophytes of gymnosperms and angiosperms and they are discharged into the atmosphere from weeds, grasses and trees. Because of wind pollination, thousands of pollen grains are liberated. While air transported pollen grains range chiefly between 10 and 50μ (micron) in size, some have been found to be as small as 5μ and as large as 100μ in diameter.

From the point of view of pollution, air pollutants are significant because of the allergic responses produced in sensitive individuals. Many people suffer from asthma or hay fever. While most victims have an uncomplicated type of hay fever in which the symptoms disappear at the end of the pollen season, some develop bronchitis, bronchial asthma, and dermatitis.

Aerosols

Aerosols refer to the dispersion of solid or liquid particles of microscopic size in gaseous media, such as dust, smoke, or mist.

An aerosol can also be defined as a colloidal system in which the dispersion medium is a gas and the disappeared phase is solid or liquid.

The term 'aerosol' is used during the time it is suspended in the air. After it has settled either by virtue of its weight, by agglomeration, or by impact on a solid or liquid surface, the term no longer applies. Thus, particulate matter is an air pollutant only when it is an aerosol. However, it is a nuisance both as an aerosol (visibility reduction) and as settled or deposited matter (soiling of surfaces, corrosions).

Aerosols differ widely in terms of particle size, particle density and their importance as pollutants. Their diameters generally range from 0.01μ or less, up to about 100μ. The following are the various aerosols.

Dust

Dust is made up of solid particles predominantly larger than those found in colloids and capable of temporary suspension in air or other gases. They do not tend to flocculate except under electrostatic forces; they also do not diffuse but settle under the influence of gravity.

Dust is produced by the crushing, grinding etc., of organic and inorganic materials. Generally they are over 20μ in diameter, although some are smaller. Fly ash from chimneys varies from 80-3μ; cement, from 150-10μ; foundry dust, from 200-1μ. Most of the dust particles settle to the ground as dust fall, but particles 5μ or smaller tend to form stable suspensions.

Smoke

Smoke consists of finely divided particles produced by incomplete combustion. It consists predominantly of carbon particles and other combustible materials. Generally the size of the particles is less than 1μ. The size of coal smoke particles range from 0.2-0.01μ and oil smoke particles from 1.0-0.03μ.

Mists

This term refers to a low concentration dispersion of liquid particles of large size. In meteorology, it means a light dispersion of minute water droplets suspended in the atmosphere. Natural mist particles formed from water vapour in the atmosphere are rather large, ranging from 500-40μ in size. The particles may coalesce.

Fog

Fog refers to visible aerosols in which the dispersed phase is liquid. Formation by condensation is usually implied. In meteorology, it refers to dispersion of water or ice in the atmosphere near the earth's surface reducing visibility to less than 1/2 km. In natural fog the size of the particles ranges from 40-1.0μ.

Fumes

These are solid particles generated by condensation from the gaseous state generally after volatilization from melted substances, and often accompanied by a chemical reaction such as oxidation. Fumes flocculate and sometimes coalesce.

Table 2 indicates the major sources of atmospheric dust.

Table 2: *Sources of Atmospheric Dust*

	Sources	*Examples*
1.	Combustion	Fuel burning (coal, wood, fuel oil) incineration (house and municipal garbage) Others (open fires, forest fires, tobacco smoking)
2.	Materials handling and processing	Loading and unloading (sand, gravel, coal, ores, lime, cement) Crushing and grinding (ores,

Contd.

		stone, cement, rocks, chemicals) Mixing and packaging (chemicals, fertilisers) Food processing (flour, cornstarch, grains) Cutting and forming (saw mills, wall board, plastics) Metallurgical (foundries, smelters) Industrial (paper, textiles manufacture)
3.	Earth-moving operations	Construction (road, buildings, dams, site clearance) Mining (blasting) Agriculture (soil filling, land preparation) Winds
4.	Miscellaneous	House cleaning, Mud road cleaning, Crop spraying, Poultry feeding Engine, exhaust

Gases

Following are the gases in air pollutants:

Sulphur Dioxide

This is one of the principal constituents of air pollutants. The main source of sulphur dioxide is the combustion of fuels, especially coal. Therefore, its concentration in the atmosphere depends upon the sulphur content of the fuel used for heating and power generation. The sulphur content of fuels varies from less than 1% for good quality anthracite to over 4% for bituminous coal. In the USA in recent years, there has been a progressive decrease in the average atmospheric concentrations of sulphur dioxide, of several cities, owing to the increased use of coal with a low sulphur content. Air pollution control measures in St Louis prohibit the use of high-sulphur coal.

Most crude petroleum products contain less than 1% sulphur: a few contain upto 5%. Refining processes tend to concentrate sulphur compounds in the heavier fractions. Fuel gases also contain sulphur, but, in small quantities.

About 80% of the sulphur in coal and nearly all that in liquid and gaseous fuels is found in flue gases in the form of sulphur dioxide. The remaining sulphur in coal is present as inorganic sulphur and thus remains in the ash. Generally, the concentration of sulphur dioxide in flue gases ranges from 0.005-0.25% and occasionally is as high as 0.4%.

Another common source of sulphur dioxide in the atmosphere, is metallurgical operations. Many ores, like zinc, copper and lead, are primarily sulphides. During the smelting of these ores, sulphur dioxide is evolved in stack concentrations of 5-10% (SO_2) but this, can be recovered in the form of sulphuric acid.

Among the miscellaneous operations releasing sulphur dioxide into the atmosphere are sulphuric acid plants and paper manufacturing plants. The quantities are usually low and therefore easily amenable to control measures.

The open burning of refuse and municipal incinerators also contribute some amount of sulphur dioxide to the atmosphere.

Hydrogen Sulphide

Hydrogen sulphide is a foul smelling gas. The sources of its natural emission include anaerobic decay processes on land, in marshes and in the oceans. Volcanoes and natural water springs emit hydrogen sulphide to some extent.

One of the major sources of Hydrogen sulphide is the Kraft pulp industry, which uses a sulphide process for manufacturing paper. The other industrial sources of hydrogen sulphide are petroleum refineries, coke-oven plants, viscose rayon plants, and some chemical operations.

Other sulphur compounds that are of interest in air pollution, principally because of their strong odour, are methyl mercaptan (CH_3 SH), dimethyl sulphide (CH_3 SCH_3) dimethyl disulphide (CH_3 SS CH_3) and their higher molecular homologs. The mercaptans are emitted in mixtures of pollutants from some pulp mills, petroleum refineries, and chemical manufacturing plants.

Hydrogen Fluoride

The major sources of fluorides are the manufacturers of phosphate fertilizers, the aluminium industry, brick plants, pottery, and ferro-enamel works. Small amounts are also emitted from other metallurgical operations, such as zinc foundries and open-hearth steel furnaces. Small amounts are also liberated in the burning of coal, which normally contains about 0.01% fluorine.

Hydrogen Fluoride is an important air contaminant even in extremely low concentrations of 0.001-0.10 ppm by volume. In these concentration levels, hydrogen fluoride is more important in terms of injury to vegetation and animals than in terms of injury to humans. The high degree of toxicity of fluorine compounds renders the control of such emissions imperative for industries manufacturing aluminium and phosphate fertilizers.

Chlorine and Hydrogen Chloride

Chlorine is found in polluted atmospheres as the element itself (chlorine) as hydrogen chloride, as chlorine containing organic compounds such as perchloroethylene and as inorganic chlorides. The last mentioned compounds are solids and hence found in particulated form; the other materials mentioned are present as gases.

The most common sources of chlorine in the atmosphere are from operations in which it is manufactured or used to produce other chemicals. Also as chlorine is used in water purification plants, in sewage plants and in swimming pools, equipment failure sometimes leads to leakage of chlorine into the atmosphere.

Hydrogen chloride is evolved in numerous industrial chemical processes, but it is so easy to recover, that little reaches the atmosphere.

The main effects of chlorine and its compounds are respiratory irritation from chlorine, corrosion by hydrogen chloride and damage to vegetation from chlorine and hydrogen chloride.

Oxides of Nitrogen

It is probable that oxides of nitrogen are the second most abundant atmospheric contaminants in many cities, ranking next to sulphur dioxide. Generally the highest concentration of nitrogen oxides in gaseous emissions occurs in effluents from industries where nitric acid is produced or used in chemical reactions. The next highest concentration is in automobile exhausts. The come effluents major from large power plants, and then to a small extent those from low heat burners and furnaces.

Out of seven oxides of nitrogen (N_2O, NO, NO_2, NO_3, N_2O_3, N_2O_4, N_2O_5), only nitric oxide and nitrogen dioxide arise from many human activities and are classified as pollutants. In atmospheric analyses they are usually reported as 'total oxides of nitrogen' or "NO_x". It is a standard practice in the chemical industry to absorb and recover significant quantities of oxides of nitrogen.

Ammonia and ammonium salts are not important air contaminants. Ammonia is an important raw material in fertilizer and organic chemical industries and in the manufacture of nitric acid by the oxidation process; its recovery is a matter of fundamental important in the economical operations of such processes.

Carbon Monoxide

Carbon Monoxide, an odourless and colourless gas has its major origin in the incomplete combustion of carbonaceous materials. It is a highly poisonous gas and is generally classified as an asphyxiant.

The chief source of carbon monoxide in the atmosphere is combustion, especially due to automobile exhausts. However, except for motor vehicles and other internal combustion engines, very little carbon monoxide is found in the gaseous emissions from properly adjusted, properly operated installations. Although certain industrial operations, such as electric and blast furnaces, some petroleum refining operations, gas manufacturing plants, and coal mines, are potential contributors of carbon monoxide to the atmosphere, automobile exhausts are by far the most important source.

Ozone

The origin of ozone that is found in the air has not been clarified, but it is likely that combustion and sunlight are involved in its production. Ozone is poisonous and smelly. It exists in great abundance under natural conditions in the upper atmosphere.

Aldehydes

These are produced by the combustion of gasoline, diesel oil, fuel oil, and natural gas. Incomplete oxidation of motor fuel and lubricating oils leads to the formation of aldehydes. Lower aldehydes may be present in the atmosphere in concentrations about as high as sulphur dioxide. They may also be formed in the air because of photochemical reactions. Formaldehyde is irritating to the eyes.

Organic Vapours

These contaminants include a large number of chemical compounds, including paraffins, olefins, acetyleness, aromatic hydrocarbons, chlorinated hydrocarbons, etc. They are produced by combustion processes (particularly in automobiles), household incinerators, and petroleum processes. It is also probable that they undergo changes in the atmosphere which contribute to the formation of smog.

Radioactive Gases

A major source of radioactive gases and particulate is the nuclear power reactor and related fuel handling facilities. The other sources are experimental accelerators, testing of nuclear bombs in the atmosphere, agricultural, industrial and medical use of radioactive isotopes. Another source of radioactive particulates and gases that is increasing in importance, is the nuclear fuel reprocessing plant.

Table 3 : *Sources of Air Pollution*

No. Class	Aerosols	Gases and Vapours
1.	Combustion processes (domestic burning, thermal power plants, cars, trucks, aeroplanes, and railways. Also refuse burning)	Dust, fume SO_2, NO_2, CO, smoke organic vapours, odours
2.	Chemical processes (paper mills, cement, fertilisers, etc.)	Dust, fume, Process-dependent mist (SO_2, CO, NH_3, NO_2, organic vapours, odours)
3.	Petroleum operations	Dust, mist SO_2, H_2S, NH_3, CO, hydro-carbons, mercaptans
4.	Metallurgical processes	Dust, fume SO_2, CO, fluorides, (aluminium refineries, organic vapours steel plants)

5.	Mineral processing	Dust, fume Process-dependent (SO_2, CO, fluorides, organic vapours)
6.	Food and feed operations	Dust, mist Odourous materials
7.	Agricultural activities (a) Crop spraying (b) Field burning	Organic phosphates, Dust, mist chlorinated hydro- Smoke, sulphur oxides, fly ash organic vapours
8.	Nuclear energy programmes (a) Fuel fabrication (b) Ore preparation (c) Bomb explosion	Dust Fluorides Iodine-131 and Argon-41 Radioactive gases (Sr-90, Cs-137, C-14, etc)

Primary and Secondary Air Pollutants

Air pollutants can also be broadly classified into two general groups - primary air pollutants and secondary air pollutants.

Primary air pollutants are those emitted directly from identifiable sources.

Examples of primary air pollutants:

1. Finer particles (less than 100m in diameter)
2. Coarse particles (greater than 100 in diameter)
3. Sulphur compounds
4. Oxides of nitrogen
5. Carbon monoxide
6. Halogen compounds
7. Organic compounds
8. Radioactive compounds

Finer aerosols include particles of metal, carbon, tar, resin, pollen, bacteria, etc.

Secondary air pollutants are those which are produced in the air by the interaction among two or more primary pollutants, or by reaction with normal atmospheric constituents, with or without photoactivation.

Examples of secondary air pollutants:

1 Ozone
2. Formaldehyde
3. PAN (proxy acetyl nitrate)
4. Photochemical smog
5. Formation of acid mists(H_2SO_4) due to reaction of sulphur dioxide and dissolved oxygen, when water droplets are present in the atmosphere.

Smog

Smog is a synchronym of two words -- smoke and fog. It can be of two types -- photochemical or coal induced.

Photochemical smog is restricted to highly motorised areas in metropolitan cities. e.g., Los Angeles. It occurs under adverse meteorological conditions when the air movement is restricted. Smog is caused by the interaction of some hydrocarbons and oxidants under the influence sunlight giving rise to dangerous peroxy acetyl nitrate (PAN). Its main constituents are nitrogen oxides, PAN, hydrocarbons, carbon monoxide and ozone. It reduces visibility, causes eye irritation, damage to vegetation and cracking of rubber.

The fog from burning coal covers urban areas at night or on cold days when the temperature is below 10°C and when calm meteorological conditions prevail, e.g., London (December 1952). This fog consists of smoke, sulphur compounds and fly ash. Prolonged exposure for smog may result in a high mortality rate especially among the elderly and those who have histories of

Chronic bronchitis, asthma, bronchopneu-monia and other lung or heart diseases.

Stationary and Mobile Sources

Emissions (air pollutants) may be classified by source, as stationary or mobile. Another method of classifying emission sources is by:

1. Point sources (large stationary sources)
2. Area sources (small stationary sources and mobile sources with indefinite routes)
3. Line sources (mobile sources with definite routes)

The following chart gives a picture of emission inventory sources classification.

Air pollution Due to Automobiles

This section discusses the contribution of automotive emissions in aggravating the air pollution menace. Various causes of the genesis and exodus of these pollutants have been identified and methods to control them have been outlined. Certain modifications in the engine design and operating variables are suggested. The possibilities of some alternatives as suitable substitutes for modern automobiles are also explored and future strategy in the Indian context has been proposed.

Emission Inventory Source Classification

Total Sources			
Stationary Sources		*Mobile sources*	
Point sources	*Area Sources*	*Line sources*	*Area sources*
1. Industrial	1. Residential	1. Highway	1. Motor vehicles
processing	heating	vehicles	light-duty
2. Power plants	coal	2. Railroad	medium-duty
3. Fuel Combustion	oil	gas	locomotives
heavy-duty			
(industrial)		3. Channel	2. Railyard
coal	2. Institutional	vessels	locomotives
oil	& commercial		3. Port vessels
gas	heating		4. Aircraft

4. Solid waste disposal	coal	(airports)
	oil	5. Miscellaneous
	gas	
5. Municipal incinerators	3. On site incineration	
Open burning		
6. Miscellaneous	4. Open burning	
	5. Evaporative losses	
	6. Miscellaneous	

The three main types of automotive vehicles being used in our country are(a) passenger cars powered by four stroke gasoline engines, (b) motor cycles, scooters and autorickshaws powered mostly by small two stroke gasoline engines, and (c) large buses and trucks powered mostly by four stroke diesel engines

Emissions from gasoline powered vehicles are generally classified as :(a) Exhaust emissions, (b) Crank-case emissions and, (c) Evaporative emissions. The amount of pollutants, that an automobile emits depends on a number of factors, including the design and operation. Of the hydrocarbons emitted by a car with no controls, the exhaust gases account for roughly 65%, evaporation from the fuel tank and carburettor for roughly 15% and blowby or crankcase emission (gases that escape around the piston rings) for about 20%. Carbon monoxide, nitrogen oxides and lead compounds are emitted almost exclusively in the exhausts gases. Table 4 gives the emissions from a typical Indian car at different conditions.

Diesel powered vehicles create relatively minor pollution problems compared to gasoline powered ones. The diesel engines exhausts only about a tenth of the amount of carbon monoxide exhausted by a gasoline engine, although its hydrocarbon emissions may be more than those of the gasoline engine. Blowby is negligible in the diesel, since the cylinders contain only air on the compression stroke. Evaporative emissions are also low because the diesel engine uses a closed injection fuel system and because the fuel is less volatile than gasoline. The major problems of diesel engines are smoke and odour.

Table 4: *Emissions from a Typical Indian Car*

Speeds (kmph)	Hydrocarbons	Carbon monoxide (ppm)	Oxides of nitrogen (%)	(ppm)
Idle		3200	3.50	75
32		2825	1.39	475
48		2475	0.28	1375
64		2175	0.18	1600
80		2000	0.18	1945

Air pollution from major Industrial Operations

Even though there are many different sources which contribute to air pollution, industries contribute a major share. There are a number of industries which are responsible for air pollution, but it is not possible to cover all of them. Table 5 shows the specific pollutants found in the emissions of some typical industrial and other sources.

Effects of air pollution on human Health

Air pollution is one of the greatest environmental evils. The air we breathe has not only life-supporting properties but also life-damaging properties. Under ideal conditions the air we inhale has a qualitative balance that maintains the well-being of man. But when the balance among the air components is disturbed, or in other words, if it is polluted, it may affect human health.

An average man breathes 22,000 times a day and takes in 16 kg of air each day. It far exceeds the consumption of food and water. It has been estimated, that a man can live for five weeks without food and five days without water, but only for five minutes without air.

All the impurities in the inhaled air do not necessarily cause harm. Depending upon the chemical nature of the pollutants, some may be harmful when present in the air in small concentrations and others only if they are present in high concentrations. The duration of exposure of the body to polluted air is also an important factor. Therefore, the prime factors affecting human health are:

1. Nature of the pollutants
2. Concentration of the pollutants
3. Duration of exposure
4. State of health of the receptor
5. Age group of the receptor

Generally speaking susceptibility to the effects of air pollution is great among infants, the elderly, and the infirm. Those with chronic diseases of the lungs or heart are thought to be at great risk. Pre-school and school children appear to be both sensitive and specifically reactive to air pollution health effects. Another point to be noted is that the effect of air pollution on human health is worst during the winter season, when pollution levels reach a climax.

An objectionable odour, visibility reduction, eye irritation or vegetation damage are useful guides to the likelihood or severity of health effects. A grey ball over a city or an industrial area can have a depressing effect and impair the enjoyment of life. There is no doubt, however, that the urgency with which steps are taken to improve air quality will depend very much on how serious the risk of ill health from air pollution is thought to be.

Table 6. *Air Pollution Episodes*

S.No.	Month and year	Place	Mortality
1.	December, 1930	Meuse Valley (Belgium)	63
2.	October, 1948	Donora (Pennsylvania)	20
3.	November, 1950	Poza Rice (Mexico)	22
4.	December, 1952	London	4000
5.	November, 1953	New York	220
6.	January, 1956	London	1000
7.	December, 1957	London	750
8.	December, 1962	London	700
9.	January, 1963	New York	300
10.	November, 1966	New York	168
11.	December, 1984	Bhopal (India)	2000

Health Effects

1. Eye irritation
2. Nose and throat irritation.
3. Irritation of the respiratory tract
4. Gases like hydrogen sulphide, ammonia and mercaptans cause odour nuisance even at low concentrations.
5 Increase in mortality rate and morbidity rate.
6. A variety of particulates particularly pollens, initiate asthmatic attacks.
7. Chronic pulmonary diseases like bronchitis and asthma, are aggravated by a high concentration of SO_2, NO_2, particulate matter and photochemical smog.
8. Carbon monoxide combines with the haemoglobin in the blood and consequently increases stress on those suffering from cardiovascular and pulmonary diseases.

9. Hydrogen fluoride causes diseases of the bone (fluorosis), and mottling of teeth.
10. Carcinogenic agents cause cancer.
11. Dust particles cause respiratory diseases. Diseases like silicosis, asbestosis, etc., result from specific dusts.
12. Certain heavy metals like lead may enter the body through the lungs and cause poisoning.

HEALTH EFFECTS OF SPECIFIC POLLUTANTS

Sulphur Dioxide

Sulphur dioxide is an irritant gas which affects the mucous membranes when inhaled. Under certain conditions, some of the air-borne sulphur dioxide gas is oxidized to sulphur trioxide. Each of these two gases in the presence of water vapour or water, forms sulphurous and sulphuric acid respectively. Sulphur trioxide is a very strong irritant, much stronger than sulphur dioxide, causing severe bronchospasms at relatively low levels of concentration.

Carbon Monoxide

Carbon monoxide has a strong affinity for combining with the hemoglobin of the blood to form carboxyhaemoglobin, COHb. This reduces the ability of the haemoglobin to carry oxygen to the body tissues. CO has about two hundred times the affinity of oxygen for attaching itself to the haemoglobin, so that low levels of CO can still result in high levels of COHb. Carbonmonoxide also effects the central nervous system. It also responsible for heart attacks and a high mortality rate.

Oxides of Nitrogen

Of the seven oxides of nitrogen known to exist in the ambient air, only two are thought to affect human health.

These are nitric oxide (NO) and nitrogen dioxide(NO_2). While some questions remain about haemoglobin reactions with oxides of nitrogen, there is no positive evidence that nitric oxide exposure is a health hazard associated with community air pollution.

Nitrogen dioxide is known to cause occupational disease. Among occupations with NO_2 hazards are the manufacture of nitric acid, exposures of farmers to silage that has high nitrate fertilisation, electric welding, and mining utilizing nitrogen compounds as explosives. It is estimated that eye and nasal irritation will be observed after exposure to about 15ppm of nitrogen dioxide and pulmonary discomfort after brief exposures to 25 ppm of nitrogen dioxide.

Hydrogen Sulphide and Mercaptans

Hydrogen sulphide is a foul smelling gas. It is well known for its rotten egg like odour. Exposures to hydrogen sulphide for short periods can result in fatigue of the sense of smell.

Other sulphur compounds that are of interest in air pollution mainly because of their strong odours are methyl mercaptan (CH_3SH) and ethyl mercaptan(C_2H_5SH). But it has been reported that at the concentrations at which they are odour nuisances, they have no other effect on human health. In fact, mercaptans are often added to natural or manufactured gas supplies so that leakage of gas will be noticed.

Ozone

Ozone is a gas that has an irritant action in the respiratory tract, reaching much deeper into the lungs than the oxides of sulphur.

Fluorides

Fluorides present in air, range from those which are

extremely irritant and corrosive like hydrogen fluoride to relatively non-reactive compounds. But, fluorine is a cumulative poison even under condition of prolonged exposure and in sub-acute concentrations.

Lead

The main source of lead in urban atmosphere is the automobile. It creates urban concentrations of inorganic lead of about 1-3mg/m^3, with high values in area of heavy traffic. Inorganic lead acts as an agent which causes a variety of human health disorders. The effects include gastro-intestinal damage, liver and kidney damage, abnormalities in fertility and pregnancy, and mental development of children gets affected.

Hydrocarbon Vapours

Some of the hydrocarbon vapours in the atmosphere have health implications. The effect of formaldehyde is primarily irritating. It is a major contributor to eye and respiratory irritation caused by Photochemical smog.

Carcinogenic Agents

Carcinogenic agents are responsible for cancer. For example, the poly-cyclic organic compound, 3, 4-benzpyrene. The origin of these compounds is in incomplete combustion of hydrocarbons and other carbonaceous materials. They are also reported to be present in exhaust discharges from I.C. engines. In addition to poly-cyclin organic compounds, it has been found that some aliphatic hydrocarbons are also carcinogenic.

Insecticides

Insecticides are not only harmful for insects but also poisonous for man,e.g., DDT (Dichloro diphenyl trichloroethane). They can affect the central nervous system and may attack other vital organs. In fact, DDT has been found in mother's milk in western countries and even in our own country. Hence, the use of DDT has been banned in the USA. The United Nations Environment Programme (UNEP) has warned against the undesirable effects of indoor spraying of DDT. It has also been reported that indoor spraying affects domestic livestock.

According to a study conducted at the Industrial Toxicology Research Centre, Lucknow (India), the accumulation of pesticides in the environment due to their growing use for agricultural purposes can also cause premature labour and abortion, due to high concentration of pesticides in the body of expectant mothers.

Radioactive Isotopes

The important radioactive isotopes that may reach ambient air are lodine 131, phospourous 32, Cobalt 60, Strontium 90, radium 226, Carbon 14, Sulphur 35, Calcium 45 and Uranium. The major sources of radioactive air pollutants are:

(a) Nuclear reactors
(b) Experimental accelerators
(c) Scientific and medical use of radioactive isotopes
(d) Agricultural and industrial use of radioactive isotopes as tracers, and
(e) Testing of nuclear bombs in the atmosphere.

The serious health effects are anaemia, Leukaemia and cancer. Radioactive isotopes also cause genetic defects and seterility, as well as embryo defects and congenital malformations. It also shortens the life span of an individual.

Effects of Air Pollution on Animals

Interest in the effects of air pollution on animals has

generally developed as a corollary to the concern about its influence on human health. Most of the information concerning the natural exposure of animals to air pollution is contained in the reports of some major air pollution disasters e.g., Donora, London and Poza Rica. Recently, considerable information has been reported from medical research laboratories which describes the results of experimental exposure of small animals to varoius air pollutants Animals used for laboratory work were mice, rabbits, rats, guinea pigs and monkeys.

Air pollution Effects on Farm Animals

The process by which farm animals get poisoned is entirely different from that by which human beings exposed to polluted atmospheres are poisoned. In case of farm animals it is a two-step process.

1. Accumulation of the air - borne contaminant in the vegetation and forage
2. Subsequent poisoning of the animals when they eat the contaminated vegetation.

In case of human beings working in polluted atmospheres in factories, the concern is for the harmful substances that are directly inhaled, whereas in the case of the farm animals, the danger obviously is not in inhaling the polluted air, but rather than ingestion of forage which has been contaminated with pollutants like fluorine from the air.

The three pollutants responsible for most livestock damage are fluorine, arsenic and lead. These pollutants originate from industrial sources or from dusting and spraying.

Fluorine

Of all farm animals, cattle and sheep are the most susceptible to fluorine toxicosis. Horses appear to be quite resistant to fluorine poisoning. Poultry are probably the most resistant to fluorine, of all farm animals.

Symptoms of Acute Fluorine Poisoning

Lack of appetite, rapid loss in weight, decline in health and vigour, lameness, periodic diarrhoea, muscular weakness and death, characterize the acute form of fluorine poisoning. It may also result in considerable increase of bone fluorine. But acute poisoning due to fluorides is unlikely in a majority of the cases.

Symptoms of Chronic Fluorine Poisoning

Fluorine is a cumulative poison under conditions of continuous exposure to sub-acute doses. Fluorine is also a protoplasmic poison. It has a marked affinity for calcium and interferes with normal calcification. Animals have been reported to be more resistant than humans to dental mottling. Cattle and sheep are the most frequently affected animals. Teeth in the process of formation are easily affected. Hence tooth symptoms are a sensitive and unique criterion of chronic fluorosis. Excessive wearing of incisor and molar teeth may occur at high levels of fluorine intake.

Bone lesions may develop at any age. A bony overgrowth may be observed on the leg bones, jaw bones and ribs. Their appearance indicates that there has been a high fluoride intake over a long period. Lameness may occur as a result of this overgrowth on the leg bones. This may be followed by stiffness of joints. Effects of fluorides on skeletal structures are usually of a permanent nature.

Symptoms of advanced fluorosis include lack of appetite, general ill-health due to malnutrition, lowered fertility, reduced milk production and growth retardation. The symptoms developed in other species such as rabbits, horses and poultry are similar to those in cattle and sheep, i.e., mottling, staining and wearing of teeth, bony overgrowths on the skeleton, stiffness and lethargy and general ill-health from malnutrition and starvation.

While diagnosis of chronic fluorine poisoning is done, one has to be careful. This is because drinking water supplies may contain unduly high levels of fluorides. Also, mineral supplements may contribute appreciable amounts of fluorine to the diet. Hence, a check of all possible sources must be made when investigating fluorosis.

Arsenic

Arsenic occurs as an impurity in many ores and in coal. It has been reported to cause poisoning of livestock near various industrial processes and smelters. Just like most industrial air contaminants, arsenic may spread over a considerable area from a stack source. Arsenic in dusts or sprays on plants can lead to poisoning of cattle.

Symptoms of Acute Arsenic Poisoning

In acute cases the symptoms are severe salivation, thirst, vomiting, uneasiness, feeble and irregular pulse and respiration. The animal stoops, lies down and gets up. There is diarrhoea, and the faeces have a garlic odour and are sometimes bloody. The ears become cold, the body trembles and develops abnormal temperature and convulsion. Death may occur in few hours or days.

Symptoms of Chronic Arsenic Poisoning

Arsenic appears to have a depressing effect upon the central nervous systems. The animal becomes dull, and exhibits a lack of appetite, with a resulting weight loss. Also, there may be chronic cough and diarrhoea may occur continuously. There may be thickening of the skin, anaemia and abortion or sterlity. Chronic poisoning can result in eventual paralysis and death.

Tolerance limit

Arsenic as well as its soluble compounds are extremely poisonous. It has been reported that sheep have been poisoned by as little as 0.25-0.50 g of arsenic daily, whereas cattle and horses may tolerate 1.3-1.9 g daily. It has also been reported from laboratory experiments, that 10 mg of arsenic per kilogram of body weight per day has produced chronic arsenic poisoning in guinea pigs. A detailed investigation of the effects of arsenic poisoning on farm animals is necessary to establish standard tolerance limits.

Lead

Lead contamination of the atmosphere takes place on account of various industrial sources such as smelters, coke ovens and other coal combustion processes. Lead is also used in dust and sprays containing lead arsenate

Symptoms of Acute Lead Poisoning

In case of acute lead poisoning the onset is sudden and the course relatively short. Prostration, staggering and inability to rise are prominent symptoms. The pulse is always fast but weak. Some animals may fall suddenly, stiffen the legs and have convulsions. There is complete loss of appetite, paralysis of the digestive tract and diarrhoea. Other nervous symptoms in cattle are grinding of the teeth and rapid chewing of the cud.

In case of horses, it may result in complete loss of appetite, nervous depression, lethargy and death.

Symptoms of Chronic Lead Poisoning

Chronic lead poisoning has been observed frequently in horses that have been grazing on forage near smelters, lead mines, and in orchards that have been sprayed. Paralysis of the muscles of the larynx and difficulty in breathing are the main symptoms. Convulsions may occur-from the paralysis of the throat, and the difficulty in breathing may be unusually severe and persistent, during and after exercise.

Tolerance Limits

Lead is a cumulative poison. Therefore, the continuous ingestion of even very small daily doses will be finally as effective as one toxic dose. Hence in case of slight contamination death can occur after many months and in case of severe contamination death can occur within 24 hours.

It has also been reported that inadequate calcium in diet increases the retention of lead in the animal body. Low - calcium diets may result in lead storage of as much as five times as compared to that found in animals that receive sufficient amounts of calcium. However, extra quantities of calcium above the amount considered sufficient, will not offer additional protection against poisoning.

Effects of Air Pollution on Plants

Air pollution has been known to have an adverse effect on plants. At first, it was only sulphur dioxide that was considered a dangerous pollutant. Now, with the advent of various pesticides and new industrial processes, the range of harmful pollutants has multiplied tremendously. Sometimes, vegetation over 150 km away from the source of the pollutant has been found to be affected.

Industrial pollution, particularly from smelters, has caused complete destruction of vegetation in some cases e.g., at Ducktown, Tennessee. The international dispute between Canada and the United States over the damage caused by the Trail, British Columbia Copper Smelter is well known. Los Angeles smog has caused widespread damage to some crops and forests in Southern California. In fact, many books and reports have been published in USA on air pollution injury to vegetation. In some cases, economics of the damage caused to crops and plants has been estimated. Compensation amounting to thousands of dollars has been paid for damage done to the crops. In our own country, there are many reports of the effect of pollutants like cement dust on plants.

Effect of Environment on Plants

The primary factor which controls gas absorption by the leaves is the degree of opening of the stomata. When the stomata are wide open, absorption is maximum and vice versa. Consequently, the same conditions that enhance the absorption of the gas (CO_2 for photosynthesis), predispose the plant to injury (by absorbing a pollutant gas like SO_2). The conditions that cause the stomata to open are high light intensity (especially in the morning hours), high relative humidity, adequate moisture supply to the roots of the plant and moderate temperatures.

Most plants close their stomata at night and are therefore much more resistant at night than in the day time. But some plants like the potato, which do not close their stomata at night are as sensitive in the dark as in the light.

Air Pollutants Affecting Plants

Sulphur dioxide	Fluoride compounds (like hydrogen fluoride)
Ozone	Chlorine
Hydrogen chloride	Nitrogen oxides (NO, NO_2, etc.)

Ammonia	Hydrogen sulphide
Hydrogen cyanide	Mercury
Ethylene	PAN (peroxy acetyl nitrate)
Herbicides (sprays of weed killers)	Smog

The above pollutants interfere with plant growth and the phenomenon of photosynthesis. Smog, dust, etc., reduce the amount of light reaching the leaf and also by clogging the stomata may reduce carbon dioxide intake to some extent and thus interfere with photosynthesis.

Forms of Damage to Leaves

Damage to leaves takes several forms

1. Necrosis	Necrosis is the killing or collapse of tissue
2. Chlorosis	Chlorosis is the loss or reduction of the green plant pigment, chlorophyll. The loss of chlorophyll usually results in a pale green or yellow pattern. Chlorosis generally indicates a deficiency of some nutrient required by the plant. In many respects, it is analogous to anaemia in animals.
3. Abscission	Leaf abscission is dropping of leaves.
4. Epinasty	Leaf epinasty is a

downward curvature of the leaf due to higher rate of growth on the upper surface.

Tissue severely injured by air pollutants often has a characteristic colour. Bleaching is associated with sulphur dioxide, yellowing with ammonia, browning with fluroide. A silvering or bronzing of the under surface of some leaves is associated with PAN injury. Phytotoxicant is the name given to plant damaging substances.

Kinds of Injury to Plants

Acute Injury

It results from short - time exposure to relatively high concentrations, such as it might occur under fumigation conditions. The effects are noted within a few hours to a few days and may result in visible markings on the leaves due to a collapse and death of cells. This leads to necrotic patterns, i.e., areas of dead tissue.

Chronic Injury

It results from long - term low level exposure and usually causes chlorosis or leaf abscission.

Growth or Yield Retardation

Here the injury is in the form of an effect on growth without visible markings (invisible injury). Usually a suppression of growth or yield occurs.

While deciding about the type of injury the plant has suffered, one must carefully distinguish between adverse effects caused by air pollution and adverse effects resulting from other factors. Plant diseases may show symptoms which are very similar to those caused by air pollution. High temperatures, poor plant care, shortage of nutrients and water may also cause an appearance which is similar to that of a plant damaged by air pollution. High temperatures, poor plant care, shortage of nutrients and water may also cause an appearance which is similar to that of a plant damaged by air pollution. It may also be due to insect damage. Therefore, while diagnosing the effects of air pollution one must consider factors such as plant disease, nutrition history, weather damage, insect damage, and the nature of pollutants in the area.

Effects of Air Pollutants on Plants

The effects of various pollutants like sulphur dioxide, ozone, fluorides, etc., on plants is shown in Table.7

Table 7: *Effects of Pollutants on Plants*

	Pollutant	Dose	Effect
1.	Sulphur dioxide	Mild	Interveinal chlorotic bleaching of leaves Severe Necrosis in interveinal areas and skeletonized leaves
2.	Ozone	Mild	Flecks on upper surfaces, premature aging and suppressed growth Severe Collapse of leaf necrosis and bleaching
3.	Fluorides	Cumulative	Necrosis at leaf tip effect
4.	Nitrogen dioxide	Mild	Suppressed growth, leaf bleaching
5.	Ethylene	Mild	Epinasty, leaf abscission
6.	PAN	Mild	Bronzing of lower leaf surface (upper surface normal), suppressed growth, Young leaves more susceptible

Effect on Sulphur Dioxide

Sulphur dioxide produces two types of injury on the leaves of plants- acute and chronic, depending on the concentration and period of exposure. The acute injury is characterized by the killing of marginal or interveinal areas of the leaf. Immediately after fumigation these areas will get a dull, water - soaked appearance, subsequently they dry up and usually bleach to an ivory colour, though some species finally assume a brown or reddish-brown colour. Chronic injury is caused by the slow, long -continued absorption of sub-lethal amounts of gas or by absorption of an amount of gas somewhat less than that necessary to cause acute injury.

Sulphur dioxide is phytotoxic in concentrations above 0.1-0.2 ppm. Below about 0.4 ppm. it tends to be oxidised in the cells as rapidly as it is absorbed, but interferences with functions such as photosynthesis is slight. Chronic injury, if any, is generally exhibited with these small concentrations.

Above about 0.4 ppm, acute injury occurs more frequently. Sometimes temporary interference with photosynthesis or 'invisible injury' can occur. However, the mechanism by which sulphur dioxide affects the plants has not yet been well understood, though various explanations have been put forth by many investigators.

Effect of Hydrogen Fluoride

Hydrogen fluoride behaves somewhat similar to sulphur dioxide except that with a few species of plants it is effective in causing lesions and interfering with photosynthesis in concentrations two or three orders less than in the case of sulphur dioxide. With most species it is up to 10 times as effective as sulphur dioxide. However, recovery of plants from the fluoride effect is much slower than from sulphur dioxide. This difference in rate of recovery is probably explained by the fact that sulphite in the leaves is rapidly oxidised to the relatively non-toxic sulphate, whereas fluorides can be removed only by the slower process of volatization or by some obscure chemical reaction. Forage may be rendered unsafe for animal feeding if more than 50 ppm of fluorine is absorbed.

Effect of Ozone

Ozone is phytotoxic in exposures of a few hours at about 0.2 ppm. Injury due to ozone is quite different from typical smog injury. The lesions are generally confined to the upper surface. The effects of ozone on plants in mild and severe doses is indicated in Table 7

Economic Effects of Air Pollution

Air pollution damage to property is very important economic aspect of pollution. In the United States of America, Great Britain, France and many other European countries, this problem has been investigated in detail and successful attempts have been made to translate observable air pollution damage into terms of economic impact. Air pollution damage to property covers a wide range - corrosion of metals, soiling and eroding of building surfaces, fading of dyed materials, rubber cracking, spoiling or destruction of vegetation, effects on animals, as well as interference with production and services. Another important economic effect of air pollution is deterioration of works of art. In our country today there is an urgent need to investigate and study this problem in detail and express the damage to property in economic terms, as very little work has been done in this direction so far.

Mechanisms of Deterioration in Polluted Atmospheres

Air pollutants cause damage to materials by five mechanisms:

1. *Abrasion*: Solid particles of sufficient size and travelling at high velocities can cause abrasive action. Also, large sharp edged particles embedded in fabrics can accelerate wear.
2. *Deposition and removal:* Solid and liquid particles deposited on a surface may not damage the material itself but it may spoil its appearance. However, the removal of these particles may cause some deterioration. Although a single washing or cleaning may not cause noticeable deterioration, frequent cleaning ultimately does.
3. *Direct chemical attack:* Some air pollutants react directly and irreversibly with materials to cause deterioration. For example, the bleaching of marble by sulphur dioxide, the tarnishing of silver by hydrogen sulphide and the etching of a metallic surface by an acid mist.
4. *Indirect Chemical attack:* Certain materials absorb some pollutants and get damaged when the pollutants undergo chemical changes. For example, sulphur dioxide absorbed by leather is converted to sulphuric acid, which deteriorates the leather.
5. *Corrosion:* the atmospheric deterioration of ferrous metals is by an electrochemical process, i.e corrosion. This due to the action of air pollutants facilitated by the presence of moisture.

Table 8 presents a generalized picture of air pollution damage to various materials.

Table 8: *Air Pollution Damage to Various Materials*

Materials	Principal air pollutants	Effects
Metals	SO_2 acid gases	Corrosion, spoilage of surface, loss of metal, tarnishing
Building materials	SO_2 acid gases,	Discolouration, leaching

Paint	SO_2, H_2S	Discolouration particulates
Textiles and	SO_2 acid gases and	Deterioration, reduced tensile strength,
Textile dyes	NO_2 Ozone	and fading
Rubber	Oxidants, ozone	Cracking, weakening
Leather	SO_2, acid gases	Disintegration, powde-red surface
Paper	SO_2, acid gases	Embrittlement
Ceramics	Acid gases	Change in surface appearance

Factors Influencing Atmospheric Deterioration

There are four prime factors which influence the rate of attack of damaging pollutants. They are moisture, temperature, sunlight, and air movement.

1. *Moisture:* The presence of moisture in the atmosphere greatly helps the process of corrosion. Without moisture in the atmosphere, there would be little atmospheric corrosion, if any, even in the most severely polluted environments. In case of sulphur dioxide and various combinations of particulates, investigations have revealed that the rate of corrosion of metals will increase as relative humidity in the air increases.
2. *Temperature:* Temperature affects the rate of the chemical reaction and consequently it affects the rate of deterioration.
3. *Sunlight:* In addition to producing damaging agents such as ozone through a series of complex photochemical reactions in the atmosphere, sunlight can cause direct deterioration of certain materials.

But in cases of fading of certain dyes or rubber cracking, damage due to direct sunlight cannot always be distinguished from that caused by ozone.

4. *Air movement:* Wind direction is an important factor to be considered in places where deterioration is caused by pollutants released from nearby factories. For example, damage to agricultural crops in the vicinity of factories. Similarly, wind speed is also an important factor in determining the impact of air pollutants on the receiving surfaces. Pollutants present in wind blowing at high speeds will have more abrasive effects and they may also be carried over long distances. Where leather bound books are stored, air movement is a critical factor when the air is contaminated by sulphur dioxide.

Effects on Art Treasures

Air pollutants have caused great damage to art objects throughout the industrialized world. Some excellent monuments of antiquity have been silently eaten away by pollutants. This can be clearly seen in many works of art sculptures and carvings on historic buildings and cathedrals. A few examples of such works of art which have been affected by air pollution both in India and abroad have been described below.

Effect on Art Treasures in India

(a) *Taj Mahal:* The Taj Mahal a miracle and marble and a historical movement is facing grave danger from pollution created by existing foundries, power house, railways yards and other industrial units. According to some environmentalists, the problem now seems to have been aggravated because of the commissioning of the Mathura Oil Refinery, about 40km from the priceless monument. Whether the monument is going to be affected by the pollutants emitted from the refinery or not has created a controversy which is now being debated. Many reports on this have appeared in noted newspapers and journals. However, it is to be hoped that certain anti-pollution safeguards built into the refinery will minimise the effect of pollutants on the monument.

According to some reports, pollutants from existing foundries power houses and railway yards are already affecting the Taj Mahal. Black and brown patches are disfiguring the shiny white surface of the Taj. Reports in newspapers also speak of the formation of gypsum, scaling and flaking effects.

Sulphur dioxide from the various sources mentioned, mixes with the atmospheric moisture and gets converted into sulphuric acid which settles down on the exterior of the Taj Mahal. The acid reacts with the marble (calcium carbonated) and corrosion and discolouration of the monument is the ultimate result.

It looks as though more money will have to be spent on protecting the Taj Mahal from the effects of pollution than it took Shah Jehan to construct it. Perhaps the immediate solution to save the Taj Mahal lies in shifting the foundries and power houses from Agra, and in effectively supervising anti-pollution measures adopted by the refinery. Further, the coal locomotives being used by the railways in that region can be replaced by diesel engines. In addition to these steps, a dense green belt of suitable tree species should be created around the Taj Mahal to absorb atmospheric pollutants as trees can act as effective air scrubbers.

(b) *Belur temple:* The renowned temple of Sri Channakeshava at Belur (Hassan district, Karnataka state) is threatened with a similar hazard. A plywood factory located close to the temple emits soot-laden fumes which get deposited on the sculptures in the temple and discolour the surface, inside and outside. This pollution effect is under investigation.

Effect on Art Preasures in Other Countries

The Parthenon in Athens, has been affected in recent years because of air pollution, Cleopatra's needle the large stone obelisk moved from Alexandria, Egypt, to London, has suffered more deterioration in the damp, smoky, acid atmosphere of London in the last 90 years than in the eariler 3000 or more years of its history. Another ancient Egyptian obelisk in New York City, shown similar decay.

Similar cases have been reported from other parts of the world. The colosseum in Rome, and the San Marco Basilica in Venice, Italy, show signs of decay, possibly due to air pollution, In France, it is reported that environmentalists have removed statues from the exterior of cathedrals and replaced them with copies. Similarly ancient temples and buildings in industrial areas of Japan are facing the hazards of atmospheric pollution. Paintings, books fabrics, antique costumes and other art objects have suffered damage.

Of course, it is impossible to make adequate substitution or compensation for the losses suffered by works of art. They are irreversible losses. It is impossible to estimate the losses in economic terms for these intangible assets. In some cases it is possible to preserve them, by applying speicial preservatives. But they are not always successful. Sensitive art objects displayed inside buildings can be placed in hermetically sealed containers. Air conditioning can also be used as a protective measure.

4 Water Pollution

Water pollution is a phenomenon that is characterized by the deterioration of the quality of land water (rivers, lakes, marshes and ground water) or sea water as a result of various human activities. Here, we will first classify various human activities that cause water pollution, discuss water pollution according to their characteristics, and then take a general view of water pollution which is currently in progress in our modern society. Human activities related with water pollution comprise various industries such as mining, agriculture, stockbreeding, fisheries, forestry, urban human activities, manufacturing industry, construction works and various tertiary industries. Then, in the last section, we will classify waste waters according to their physical and chemical properties. Water pollution is any physical or chemical change in water that can adversely affect organisms. It is a global problem, affecting both the industrialized and the developing nations. The water pollution problems in the rich and the poor nations, however, are quite different in many respects. Heat, toxic metals, acids, sediment, animal and human wastes, and synthetic organic compounds foul the waterways of developed nations. Human and animal wastes, sediment, and pathogenic organisms head the list in the nonindustrialized nations. In these countries, unsanitary water and malnutrition account for most of the illness and death. Water has many different uses, each requiring different levels of purity. For example, water from the Ohio River may be clean enough to wash steel but may be toxic to fish and wildlife: water suitable for boating or fishing may be unsuitable for swimming : water safe to swim in may be too polluted to drink.

Like air pollutants, water pollutants come from numerous natural and anthropogenic sources. Likewise, water pollutants produced in one nation may flow into others, creating complex international control problems that may take decades to solve.

Major water Pollutants

Following are the eight major types of water pollutants:

1. Oxygen-demanding wastes (domestic sewage, animal manure, and some industrial wastes)
2. Disease-causing agents (bacteria, parasites, and viruses)
3. Inorganic chemicals and minerals (acids, salts, and toxic metals)
4. Organic chemicals (pesticides, plastics, detergents, industrial wastes, and oil)
5. Plant nutrients (nitrates and phosphates)
6. Sediments (soils, silt, and other solids from land erosion)
7. Radioactive substances
8. Heat (from industrial and power plant cooling water)

Although sedimentation is the major water pollutant in terms of quantity, other water pollutants can pose serious threats to plants and animals (including humans) when present in minute quantities. Some organisms have low tolerance levels to chemicals such as oil, some pesticides, and lead and mercury compounds. Other organisms with higher tolerance levels can be affected when the levels of some pollutants are biologically concentrated in food chains.

Water pollutants can be classified as degradable (either rapidly or slowly), or nondegradable *Rapidly degradable* (non persistent) pollutants can be broken down fairly quickly by natural chemical cycling processes as long as the pollutants do not overload the system. They include domestic sewage, other oxygen-demanding wastes, plant nutrients, and some synthetic organic chemicals. To control these pollutants we need to prevent overload. Normally these pollutants are degraded into harmless or less harmful forms, but sometimes they are converted into more harmful forms. For example, microorganisms in acidic waters can convert moderately harmful mercury metal and inorganic mercury compounds into very toxic organic methyl mercury.

Slowly degradable (persistent) *pollutants* remain for long periods of time but are eventually broken down or reduced to harmless levels by natural processes. Examples include some radioisotopes and many synthetic organic chemicals, such as DDT, PCBs, phenols, and older-type detergents. Some of these substances, such as DDT, may be magnified up the food chain to levels that are harmful for wildlife and possibly for humans because most of these pollutants are extremely difficult to remove by waste treatment methods, control measures must emphasize (1) preventing or minimizing their entry into the environment, (2) long-term storage until levels are safe, and (3) research to determine which of these pollutants are dangerous and capable of being magnified up the food chain. *Nondegradable pollutants* are not broken down by natural purifying processes. They include some metals (such as mercury, lead, and arsenic), some salts of metals, sediments, some synthetic organic compounds (such as plastics), and some bacteria and viruses. These pollutants must be controlled either by removing them through waste treatment or by preventing them from entering the environment.

Sources of Water Pollution

Water pollution can result from natural runoff; dissolved chemicals in water that percolates through the soil; and human sources, such as agriculture, mining, construction, industry, homes, and businesses. It is useful to classify pollution sources as either point sources or nonpoint sources. Point sources involve discharge or wastes from identifiable points. They include (1) sewage treatment plants (which remove some but not all pollutants) (2) storm water runoff from combined storm and sanitary sewer lines in urban areas; (3) industrial plants; and (4) animal feedlots where large numbers (sometimes 10,000 or more) of animals are scientifically fed in tightly restricted quarters before slaughter. Nonpoint sources involve the diffuse discharge of wastes from land runoff, atmospheric washout, and sources that are difficult to identify and control. They include (1) runoff of sediment from natural and human caused forest fires, construction, logging, and farming; (2) runoff of chemical fertilizers, pesticides, and saline irrigation water from croplands; (3) urban storm water runoff; (4) drainage of acids, minerals, and sediments from active and abandoned mines; and (5) spills of oil and other hazardous materials. Nonpoint source pollution is not recognized as a major problem, since the sources are widely spread out, difficult to identify, and hard to control.

Water Pollution by Agriculture

Agriculture has been a victim of water pollution in many instances, but sometimes, it is also responsible for polluting water. Water pollution caused by agriculture is mainly an outcome of fertilizers and agricultural chemicals such as insecticides and herbicides.

(1) Water Pollution by Fertilizers

Fertilizers given to crops are not always fully consumed, and part of it remains in the soil by being absorbed by soil colloid, and influence the quality of underground, river and sea waters when it is dissolved. Chemical fertilizers mainly consists of relatively simple compounds of nitrogen, phosphorus and potassium which are the nutritive elements of plants. Their runoff coefficients vary depending on the solubility of the fertilizer itself, the rate of absorption by plants, rate of absorption by the soil and rate of decomposition. It is generally believed that the runoff coefficient of nitrogen is 5-30%, and that of phosphorus is 0.5-5%. Especially, nitrogen having a high runoff coefficient tends to cause the eutrophication of lakes and enclosed sea areas.

Fertilizers as the sources of water pollution are called "non-point sources" in contrast to "point sources" such as sewage and industrial waste waters. The non-point sources are not only hard to eliminate, but also hard to measure their quantity. Organic fertilizers such as compost, farmyard manure, human waste and fish meal produce nitrogen and phosphorus when they are decomposed, and also become the sources of pollution. These organic fertilizers do not contain, in general, any toxic substances.

(2) Water Pollution by Agricultural Chemicals

Agricultural chemicals are considered as a source of peculiar pollution which cannot be overlooked. Agricultural chemicals comprise a variety of chemicals such as insecticides, sterilizers and weed killers, and have played important roles in the operation of modern agriculture, but they have also caused serious environmental pollution.

The characteristics of agricultural chemicals as water-polluting substances lie in that (1) almost all of them are special compounds not occurring at all in nature, and many of the behaviours and changes of such chemicals at the time when they are discharged into the natural environment are known; (2) many of their influences on the human body because of their inherent characteristic; (3) when they are ingested into the human body, they tend to accumulate or concentrate in it; (4) they are available in a variety of kinds, and new products are developed constantly; (5) some of them are extremely stable as chemicals, and remain in nature or in the human body for a long time thereby eventually afflicting the human body; (6) it is common to all such chemicals to contain additives besides the main constituents, and the additives can contribute to affect the environment though they do not have a strong toxicity; (7) many of them require a special method of analysis.

Of the agricultural chemicals developed in the past, many of them are prohibited from being used because of their violent toxicity. Nevertheless, those remaining in the environment are still factors that contribute to environmental pollution.

In addition to above-mentioned pollutions, there are instances of water pollution caused by indirect effect of agricultural activities. For example, the change in the condition of the ground surface by reclamation of waste land, deforestation, supply of the soil to low land, and the alteration of the natural water system for irrigation purposes can be considered indirect causes of water pollution.

Water Pollution by Mining Industry

In many countries, water pollution related to the mining industry can be traced back to olden times compared with those caused by other industries.

The characteristics of waste water from mining industries varies greatly depending on the kind of mine. For instance.

(i) Metal Mining Industry

Many kinds of waste waters are discharged. For example, mine water, ore dressing water, waste water from smelters, etc. In general, it is strongly acidic (low pH), and contains metal irons, sulphuric acid and suspended solids of high concentrations.

(ii) Coal Mining Industry

Waste water from coal mines mainly consists of mine water, coal dressing water, exudative water from coal-waste heap etc.. and contains a large quantity of suspended solids. In general, the scales of operation of coal mines is much larger than that of other mining industries, and thus the coal mining industry exerts a large influence upon the natural environment.

(iii) Sulphur Mine

In general, waste water from sulphur mines contains large quantities of sulphuric acid and iron and has a low pH value. Almost all sulphur mines are deserted; so the administration of deserted mines becomes a social and environmental problem.

(iv) Petroleum and natural Gas Drilling Industries

Different from other mining industries, these industries are very often located in plains. Waste water generally consists of welled-up mine water and waste water incidental to the drilling operations. It contains large quantities of salt and organic matters such as mineral oil.

(v) Stone-quarrying and Clay-supply Industries

These industries supply lime stones, building stones, porcelain clay etc. Waste water from the sites of these industries generally contains large quantities of suspended solids.

The characteristics of industrial pollution common to the above-mentioned mining industries can be summarized as follows:

(i) In many instances, the development of a mine itself changes its natural environment artificially, and such a change first affects the quality of water of the environment even if any kind of waste water is not discharged from mining activities.

(ii) In areas where mines can be developed, very often, even before the development of the mine, the quality of river water or underground water is different from that of other areas. For instance, river water or underground water in such areas contain special metals of high concentrations and is strongly acid. Thus, it is rather difficult to measure the direct influence of mining activities on the quality of water. And also, geological features and the ways of land use are usually very peculiar in mining areas. These conditions should be taken into consideration prior to surveying the quality of water.

(iii) Mining activities generate large quantities of solid wastes in most cases, and such wastes from large heaps. Such wastes not only pollutes water by exudation of toxic substances, but also, the rapid generation of suspended solids in water when the heap is washed away by a heavy rainfall or a flood.

Water Pollution by Stockbreeding and Fisheries

1. When cattle or horses are fed on a big pasture, there is little possibility of the livestock causing water

pollution because their excreta are spread widely all over the pasture. However, when livestock is fed intensively in a limited area it causes a number of problems. In the case of the stockbreeding industry, the excreta and leftover feed on livestock are the sources of water pollution. Water pollution caused by stockbreeding has the following characteristics.

(i) Degree of pollution varies largely depending on the location of the stockbreeding farm (whether in suburban or country area), system and scale of operation and the level of breeding technique.

(ii) Composition and quantity of excreta of livestock varies depending on the type, quantity and methods of feeding and watering.

(iii) Besides water pollution, the stockbreeding industry also generates an offensive odor, flies, noise and so on.

2. Besides stockbreeding farms, waste water from slaughterhouses is a source of water pollution which should not be overlooked.

3. In most instances, fisheries has been the victim of water pollution. But in recent years, as the number of fish farms increase, pollution of water areas surrounding fish farms caused by leftover feed and excreta of fish is becoming serious.

Water Pollution by Urban Activity

In a sense, the existence of human beings itself is a source of water pollution. In primitive ages, the purifying capacity of nature was much greater than the rate of water pollution caused by human living. But as time passed, men began to settle near water areas for the convenience of living; then they developed various industries which provided them with the foundation of civilized life. The emergence of various industries coupled with the development of civilization has constantly increased the quantity of waste water containing pollutants. On the other hand, the increasing concentration of the population has also invited the concentration of pollution sources. As a result, the rate of environmental pollution has exceeded the rate of natural purification.

If all the urban activities of human beings related with water pollution are viewed as a metabolic system, it can be outlined as follows:

	Waste water
Water	Human waste and municipal waste
Energies	Waste gas and waste heat

Substances Municipal wastes such as waste water, human waste, household waste, waste gas and waste heat are the causes of water pollution.

The type of water pollution in a city varies largely depending on the characteristics of the city, life style of the inhabitants, degree of civilization, degree of development of sewage treatment plants or sewerage systems. Although, it is very difficult to investigate these conditions in detail, they can be estimated by investigating the following:

(i) Population of the city: This is the most essential datum to be obtained first.

(ii) Consumption of water and quantity of waste water: where a city water system is well established, it is relatively easy to obtain data concerning the consumption of water. For example, the consumption of city water in a modern civilized city is said to be about 500 liter/man/day. However, it is usually difficult to know the quantity of waste water even if it is only an estimate.

(iii) Condition of sewage treatment: This can be measured in terms of the availability of sewage treatment plants and sewage works including their capacities and efficiency.

(iv) Condition of municipal or public facilities and private facilities: In a large city, there are a number of public facilities and various other facilities, and all these facilities can be the sources of water pollution, such as: (public facilities) Government and municipal offices, schools, hospitals, public bath houses, public lavatories, markets, ports, power plants, city gas facilities, sewage treatment plants. etc.

(Non-public facilities) Factories, stores, offices, gas stations, restaurants, bars, hotels, etc.

(v) Condition of sewage: Municipal sewage mainly consists of a mixture of human waste, household sewage and wastes from other facilities, and its condition varies depending on the characteristics of the city, degree of civilization, precipitation and the condition of road surface.

The quantity and condition of waste water resulting from human activity, mainly consisting of household waste water, generally vary depending on the time in a day.; however, the "product of the concentration multiplied by the quantity" tends to remain within a relatively fixed range. Such a value is defined as "pollution loading amount", and is used as the basic data for the sewage treatment planning.

Table 1: Standard values of BOD, SS and pollutant load of sewage and human waste

Type of waste water	*Load, equivalent, concentra-tion, etc. of BOD and SS*
Human waste from flush toilets (including those	(1) BOD equivalent per person per day: 13 g

from septic tanks for sewage)	(BOD: 260 mg/) (2) SS equivalent per person per day: 26 g (SS: 520 mg/1) (3) Flushing. water per person per day: 50/
Household waste water (not Including rainwater and special waste water)	(1) BOD equivalent per person per day: 40 g (BOD: 200 mg/1) (2) SS equivalent per person per day: 50 g (SS: 250 mg/1) (3) Quantity of waste water per person per day: 200/
Other typical waste waters	(1) Average BOD of waste water from office building; about 80 mg/1 (2) Average BOD of waste water from hotels about 100 mg/1 *Contd.* (3) Average BOD of waste water from hospitals about 100 mg/1 (4) Average BOD of waste water from restaurants and catering facilities: about 400 mg/1

The most conspicuous characteristics of municipal sewage are that the organic content (indicates as BOD) is extremely high; the contents of ammoniacal and organic nitrogen, phosphates and sulfides are considerably high; and normally, hazardous substances are not contained

Industrial waste water tends to contain hazardous substances; therefore, its discharge into the municipal sewerage system should be strictly controlled. This is necessary because ordinary sewage treatment plants not only do not have an equipment to remove hazardous substances, but also there is a possibility of getting its activated sludge processing function damaged by such hazardous substances.

The modern sewage treatment technique is capable of removing organic pollutants very efficiently, but it is difficult economically to re-move nitrogen and phosphorus contents efficiently with existing facilities. For example, only about 30% of the nitrogen and phosphorus contents can be removed through the currently employed two-step treatment method, which mainly applies the activated sludge method, and the remainder of such contents causes the eutrophication of water areas.

Anyhow, all of the above mentioned techniques are possible where the municipal sewerage system and sewage treatment plants are well established, and the quantity and quality of waste water can be measured effectively. However, where sewage treatment plants are not available, or where waste water is discharged directly into the hydrosphere without passing a sewage treatment plant, it is almost impossible to measure the quality and quantity of waste water.

In case waste water is directly discharged into the hydrosphere, and if the pathogenic bacteria of an infectious disease is contained in the human waste, such bacteria would be discharged into the hydrosphere in active condition thereby eventually spreading the infections disease through the water system. The degree of such a danger various depending on the quality of sanitation facilities of individual cities.

Water Pollution by Manufacturing Industry

Characteristics of water pollution by Manufacturing Industry

Of water pollutions with which human activities are concerned, the manufacturing industry is the largest in scale, and most diversified. The characteristics of the water pollution of manufacturing industry can be summarized as follows:

1. Unlike the mining industry and agriculture, the discharging condition of waste water from manufacturing industries is relatively simple because it is rarely affected by complicated conditions of the natural environment. For example, waste water of the manufacturing industry is discharged concentratively from a fixed outlet, which is called "point source pollution". Besides, waste water from the manufacturing industry is generated as a result of the production activity which is 100% artificial, and thus the manufacturing industry can be considered as the source of water pollution which is most easy to catch and control in this point.
2. Sources of water pollution from manufacturing industry are relatively easy trace back through scientific method. But their control is not always easy because it involves complex social and political problems. Furthermore, tracing such point sources of pollution is not always easy because the interests of individual manufacturing companies can be affected by pollution control measures.
3. In contrast to the relatively simple form of discharging process, the composition and quantity of waste water vary widely depending on the type and scale of individual factories, change from time to time, and its influence also takes a variety of forms.
4. The influence of waste water on the environment varies depending on the geographical location of the factory. For example

(i) In the case of an industrial conglomerate or large industrial complex: The influence of waste water on the environment is extremely large, but as far as such manufacturing facilities are concentrated in one area based on the results of careful environmental assessment, waste water from these facilities can be controlled in the most ideal form;

(ii) In case of factories are located in commercial or residential areas of a city: The location of this type is primarily undesirable. Actually,however, there are a number of such cases because of historical reasons, or geographical advantages concerning the purchase of raw materials, transportation of products, or access to the labour market. In Japan, locations within easy access to the source of industrial water are decreasing in these areas. As a result, factories that recycle water are increasing; small scale manufacturing companies establish cooperative associations and are promoting the construction of joint waste water treatment facilities;

(iii) Location of the manufacturing factory is whether mountainous area, plain or coastal region is also an important factor to be considered in assessing the influence of waste water on the environment. Especially, in the case of a water-consuming type of manufacturing industry, its location has to be decided depending on accessibility to an industrial water source. On the other hand, the condition of the river or the sea into which waste water of the factory has to be discharged also has a great influence on the form and type of environmental pollution.

Characteristics of waste water of Main Industries

As discussed in the forgoing, the composition of industrial waste water is complex, and it varies largely according to the difference in the manufacturing process even among the same type of manufacturing factories. Furthermore, the manufacturing industry is constantly subject to the impact to technological innovations which occur at increasingly short intervals. For example, in

Japan, the technique for recycling of waste water is making rapid progress. Manufacturing companies were compelled to do so due to the increasingly strict pollution control requirements and the increasing shortage of industrial water. As a result, manufacturing companies are also compelled to alter conventional manufacturing processes. It is rather difficult to enumerate the characteristics of waste water of each industry General characteristics of waste water according to the type of manufacturing industry are as follows:

(1) Food Processing Industry

A common characteristic of waste water from all types of food processing industries is the high concentration of organic content. Another common characteristic is the high concentration of solid, semisolid and suspended solid contents in waste water discharged from raw material treatment processes. The quality and quantity of waste water vary depending largely on the season mainly due to the variation of raw materials to be processed. Materials contained in waste water can be divided roughly into carbohydrate, protein, fat and a mixture of them. In many instances, waste water contains oil, nitrogen and phosphorous. In the case of starch producing, super refining and brewing industries, the BOD level of waste water is especially high.

(2) Textile Industry

There is much difference in the content of waste water between natural and chemical textile industries.Waste water from the wool industry contains high concentrations of BOD, fat and alkali. Waste water from dyeing discharges various pollutants such as dyes, auxilliary additives and other chemicals.

(3) Paper and Pulp Making Industries

These are typical water-consuming industries and the

levels of COD and suspended solids in waste water are extremely high. Especially, waste water from the digesting process contains all components of wood except cellulose and has high COD and dense colour, so that it is necessary to collect and treat most of the content to recover. A large quantity of waste paper fibre is also discharged as supended solids together with the waste water.

(4) Petroleum Refining Industry

The main component of pollutants is spilled oil; but waste water also contains hydrogen sulfide, other sulfides, ammonia, mercaptan, and phenols,etc.

(5) Petrochemical Industry

The waste water contains hydrocarbons, various organic compounds and catalysts, and its content varies largely depending on the type of product.

(6) Iron and Steel Making Industry

This industry is also a water-consuming industry. Waste water from the cooling and cleaning processes for coke furnace gas contains ammonia, cyanide, phenols, etc. Besides, there is waste water from the dust collecting process of each furnace which contains suspended solids (coke, dust and ore), and that from the pickling process which contains acid, iron and oil.

(7) Electroplating Industry

Many of the raw materials and chemicals used in this industry contain toxic substances, and thus the waste water contains various heavy metals (cadmium, zinc, copper etc.), cyanide, hexavalent chromium, acids and alkalis.

(8) Leather Industry

The levels of BOD is high in waste water which comes

from tannin used for tanning and raws. It also contains chromium used for tanning and lime for pretreatment, suspended solids and dyes.

(9) Steam and Nuclear Power Plants

These power plants discharge large quantities of effluent water from the cooling process. In Japan, all the large power plants are located close to the sea, and these plants are using sea water for cooling. The temperature of waste cooling water is normally high, and such waste water can cause thermal pollution or the elevation of temperature to the vicinity of discharing outlet. No other pollution is considered.

Water Pollution by Other Human Activities

Aside from the various human activities already discussed in previous sections, there are other types of human activities which also cause water pollution, and such activities sometimes become social problems as sources of environmental pollution. These types of human activities are as follows:

(1) Construction work

Construction works, especially those large-scale land development projects or forestry conservation and flood control projects aimed at preventing natural disasters, sometimes can cause water pollution: Water pollution caused by such construction projects can be divided into those occurring incidental to the progress of construction works and those occurring due to the change in natural conditions as a result of the construction works. The latter is more important than the former. Construction works for a dam, river improvement (including those for the alteration of basins), port facilities, creation and reclamation of land and gravel gathering from river-bed are enumerated as examples of the above-mentioned causes.

(2) Various tertiary industries

Various tertiary industries can be considered to cause water pollution such as cleaning and laundry businesses, public bath houses, hotels, printing business, photographic film development businesses, hospitals, laboratories or research institutes, tourist businesses, transportation facilities and so forth. An increase in the load of municipal waste water as a result of the increase in the load of municipal waste water as a result of the increase in the working population of cities also has become one of the factors directly related to water pollution. Of various transportations, ships, especially tankers which discharge waste cleaning water, have come to the fore as sources of water pollution in coastal regions.

(3) Errors, accidents and disasters

Accidental damages to public sewage works or drainage of industrial effluent, as well as poor planning of sewage treatment plants, and erroneous operations of these facilities can also cause water pollution. Speaking of water pollution caused by accidents, pollution of the ocean due to oil resulting from the collision of tankers on the high seas has become an object of public attention in recent years.

By now we have taken a look at conditions of water pollution caused by human activities classifying by the type of activity. So as look at water pollution from a different angle, we may classify and explain it from the standpoint of its physical and chemical characteristics.

(1) Suspended solids (Abbreviated as "SS")

The causes what is called the "turbidity of water". In general, suspended solids are defined as those having a diameter of over 1u. Waste water from cement manufacturing industry and some of the mines can be enumerated as examples of waste water containing inorganic SS. Municipal sewage and waste water from kraft pulp factories are typical examples of those containing organic SS.

(2) Organic matter

Organic matter is usually measured as BOD or COD. Organic matter consumes the dissolved oxygen in water, and eventually destroys the ecosystem of the hydrosphere. This is the main content of not only municipal waste water, but also industrial waste water such as those from food processing, chemical, pulp and leather industries.

(3) Inorganic matter

Inorganic matter comprises acids, alkalis, various salts, metals, etc. Waste water containing such substances are those from mines, chemical, pulp, electroplating and other industries.

(4) Inorganic and organic toxic substances

Toxic substances are defined as chemical substances even traces of which adversely affect either the human body or the living environment.

(a) Inorganic toxic substances

Comprises metallic substances such as cadmium, lead, hexavalent chromium, mercury etc.., and non-metallic ones such as cyanide and arsenic; and are contained in various industrial and mining waste water.

(b) Organic toxic substances

The modern chemical industry has created a number of new synthetic substances and their special characteristics are utilized to develop new agricultural chemicals and medicines, etc. However, these synthetic substances adversely affect the environment when they are discharged into it. Many of these substances consists of organic compounds containing phosphorus, chlorine, mercury, etc.

(5) Oil and fat (hexane Extractable Component)

Comprise vegetable oils, animals bats and mineral oils. Oils and fats are found in municipal waste water and in waste water from food, oil and fat processing industries, etc. While mineral oils are found in the waste water from oil drilling, petroleum refining, petrochemical, iron and steel making and other industries. They are also found in waste water from ships. In Japanese standard method the content of oils and fats is measured quantitatively by extracting them with hexane and is called "hexane extractable component".

(6) Substances contributing to Eutrophication (Nitrogen Compounds and Phosphates)

Salts such as nitrogen and phosphoric acid compounds NH_4,NO_2, and PO_4) exist in small quantities even in natural clean water, and these substances are ingested by organisms as nutrient salts; but they also contribute to the eutrophication of water areas when their quantities exceed certain levels these substances are discharged contained in municipal waste water, surplus of fertilizers, some industrial waste water and synthetic detergents.

(7) Sulfur compounds

Hydrogen sulfied (H_2S) is a typical sulfide that causes water pollution, and is formed in municipal waste water, human waste, waste water from the digesting process of pulp and waste water from gas plants. Hydrogen sulfide gives an offensive odor, and corrodes the facilities and buildings located near the passage of waste water because of its strong reducing power. The content of hydrogen sulfide indicates anaerobic index of the water area. In the volcanic zone, hydrogen sulfide frequently occurs naturally. Sulfur oxides are main pollutants in air and fall to the earth as "acid rain".

(8) Coloring and odor-giving substances

Even if not directly harmful to human health, coloring or odor-giving substances in water areas can seriously influence the living environment of people.

Such substances are found in the waste water from dyeing, chemical, agricultural product processing, marine product processing and other industries.

(9) Coliform group

Colon bacillus itself does not have any pathogenic nature, but it is used as an important index to measure pollution by human waste.

Water pollution due to human excreta is caused mainly by the lack of a proper municipal sewerage system and is important since there is a possibility to contain pathogenic organisms.

(10) Radioactive substances

Waste water from research institutes or hospitals that use isotopes, nuclear fuel treatment plants and nuclear power plants are considered to include radioactive substances.

(11) Warm and hot waste waters

Warm or hot waste water consisting of use cooling water discharged from factories and steam or nuclear power plants can significantly influence the water area as part of the living environment when it is discharged in large quantities. However, when the temperature of such waste water is controlled properly it can be utilized for fish farming or district heating.

Table 2 shows the major sources activities that cause water pollution.

Table 2: Major causes and sources of water pollution

Cause	*Source (industries and others)*
BOD Content	Food processing industry (canned goods, meat and dairy product (organic matter) processing, marine product processing. Starch process-ing), brewery, sugar making, wool processing (wool washing), pulp and paper industry, leather industry, municipal sewerage night soil treatment plant. Suspended solids (Organic SS) Food processing, paper and leather industries, coal (SS) washing, municipal sewerage. (Inorganic SS) Ceramic industry, gravel yard, quarrying industry, mining industry. Oils and fats (Oils, Food processing, petrochemi-cal, iron and steel making, metal fats, miniral Oils) finishing, wool processing (Wool washing) industries. Acids or Alkalis Mining, paper making, pulp, electroplating, metallurgical, chemical, textile and leather industries. Ammonia Gas and coke plants Phenols Gas and coke plants, synthetic resin and other organic chemical industries.
Sulphides	Pulp and gas industries, Sewage treatment plants, dyeing and leather industries.
Color	Dyestuff manufacturing, dyeing, paint manufacturing, leather, chemical industries, night soil treatment plants.
Odor	Chemical, pulp, marine product processing, meat processing, sugar making, leather industries, sewage and night soil treatment plants.
Detergent	(Surface Municipal sewerage, textile industry, laundry. active agent)
High-temperature	Steam and nuclear power water plants.

Eutrophication

Phosphorus and nitrogen compounds (phosphates, ammonium salts, nitrites and nitrates, etc) are called nutrient salts which are essential to the growth of phytoplanktons and algae participating in the primary production. However, if they are excessively supplied together with organic substances, the production activities become too vigorous due to overnutrition, ultimately changing the aquatic ecosystem. This phenomenon is called eutrophication. Eutrophication of lakes is not an extraordinary phenomenon but one which occurs and slowly proceeds even in natural not polluted ecosystem. However, eutrophication in polluted sea and lakes progresses very quickly due to rapid increase of organic substances and nutrient salts, causing unusual growth of specific organisms and thus destroying the normal aquatic ecosystem. This type of eutrophication is often observed in closed water areas such as lakes and deeply recessed bays; the results is colouring and smelling of the drinking water taken from the lakes, lowered clarity of lake water, or the occurrence of the so-called "red tide" in sea and lakes as a visible and conspicuous phenomenon. The red tide is caused by unusual propagation of certain kinds of planktons at such a terrific speed as to change to colour of water.

Consequently, many environmental trouble occur such as death of fishes due to clogging of gills with planktons, oxygen deficiency over the polluted area, and generation of poisonous substances.

In Japan, this phenomenon has often been observed in the inland Sea and other closed water areas. The true causes have not yet been determined exactly, but it is generally accepted that phosphorus and nitrogen compounds contained in urban and industrial waste waters and playing an important role.

Contamination of bottom sediments

River, lake and sea bottoms are usually covered with a certain depth of muddy sediments. These sediments contain remains and excreta of aquatic organisms, other than the earth and sand carried over by rivers. Suspended solids in water tend to sediment remarkably in lakes, slow-flowing rivers, estuaries where the river stream is stagnated by tidal currents, and in closed sea areas.

Cases are also reported where some highly poisonous heavy metals and sparingly soluble organic substances, which had settled on the bottom of water and fixed as sediment over a long period, where dissolved again or raised under some conditions, causing serious hazards. A typical example is mercury in the bottom sediment of the Minamata Bay. It was about twenty years ago when mercury caused the Minamata disease, but mercury hazard are still continuing Minamata disease, but mercury hazards are still continuing even after the discharge of mercury from industrial waste water completely stopped.

At the bottoms of lakes and closed areas with high organic loads, dissolved oxygen is consumed rapidly during the putrefaction of organic substances contained in the bottom mud. If vertical mixing of water is insufficient due to stratification, oxygen dissolved on the surface of water from the atmospheric air cannot reach the bottom. In addition, poor clarity of water weakens sufficient penetration of the sun beam and significantly suppresses photosynthetic reactions in the bottom water layer. Under these conditions, the dissolved oxygen in the bottom water mass will decrease, ultimately leading to the zero-oxygen state. As a result, fishes in that area will be seriously, affected and the surrounding ecosystem may be materially destroyed.

Other causes are also known where elusion of phosphorus and nitrogen from bottom sediments in closed sea areas may be responsible for eutrophication and the resultant occurrence of red tide. Water bottoms often provide favourable spawning grounds for fishes; anaerobic conditions of bottom sediments destroy hatching in many causes.

A survey of bottom sediments provides the environmental information of the neighboring water area in the past over a considerably long period. Thus, it is possible to obtain new findings on the progress of pollution not expected from water quality investigations.

Oil Pollution in oceans

Modern civilization relies in a large measure on petroleum in all aspects. A huge amount of crude oil is carried by supertankers on oceans throughout the world, stored in tanks at terminal stations, and processed into various petroleum products at petrochemical complexes. Many terminal stations and petrochemical complexes are located in coastal zones. Probably a considerable amount of oil is always flowing into the ocean. Including slight leaks from various ground facilities, leakage from tankers during loading and unloading, and discharge of waste oil used on land. Besides, large scale oil spills due to maritime accidents have often occurred.

Oil spilt out to the sea tends to-spread on the sea surface as a thin layer. Volatile components gradually evaporate and some of them dissolve into water (benzene, for example, dissolves in seawater to a concentration of about 820 ppm at ordinary temperature). The oil membrane thus formed on the sea surface disturbs normal exchange of gases between the seawater and the atmospheric air, and also interrupts the penetration of sun beams to retard the growth of planktons.

The oil gradually becomes thick as the volatile components evaporate; the resultant viscous heavy oil is in or on the seawater, or gather shellfishes and sand on the sea bottom to form oil balls, which drift over long distance and coast shore. The spilt oil is degraded by some bacteria though very slowly. However, if it drifts faster than it decomposes, pollution will proceed. Heavy oil sometimes sticks to the bodies of seafowls and fishes; particularly, larvas are seriously damaged by oil deposition. Although not so critical, causes are often encountered when fishes cannot be used for food because of the smell of heavy oil.

5 HIMALAYA

Everyday we have new warnings and new declaration on the need to protect our mountains. So we have the "Declarations of Kathmandu of 1982," and we have the "These de Biella of 1988," ("Mountaineers of the whole world unite in defence of mountains") written by all the great names in mountaineering : Hillary, Messner, Bonnington, and the rest.

And everyday we have zealous and strident efforts to attract more people to unspoilt regions. M.Krishnan, the wildlife conservationist, recently wrote of his disillusionment with "the chasm of insincerity and futility between proclaimed policy and the invariable concessions to all popular depletive factors." What is true of wildlife is true of mountains.

The Swiss Alps, the original mountain playgrounds of the world, are a prototype. For a hundred and fifty years, there was a harmonious balance between the players and the playground. Mountaineers, walkers, campers, Skiers, artists, lovers of nature, had their pleasure without humiliating the dignity of the mountains. But that was in the dead past.

Climbing huts have been overwhelmed by numbers. The Swiss Alpine Club has 154 climbing huts from the Britannia with bunks for 120 and catering to over 8000 overnight visitors a year, to the Silberhorn with accommodation for only 12 with about 80 overnight visitors. Most of the huts have had to be enlarged to accommodate those who are enticed by easier access today (and cheap accommodation). Garbage and toilets were not a problem when visitors were limited; they are now. They certainly are in the Himalayas and though the numbers of visitors may be smaller, the problem is greater.

It is skiing, however, that has brought the greater pleasure and wrought the greatest damage to the Alps. It attracts millions and is a multi-billion dollar industry. In Switzerland there are some 500 major ropeways and chair lifts, and well over a 1000 ski-tows But it is now universally recognized that saturation has been reached indeed exceeded—and some 15 to 20 requests for new mechanical lifts are rejected every year. When the winter of 1989-1990 turned out to be catastrophically short of snow, it was an economic disaster for hotels, ski ships, manufactures of equipment, ski monitors, operators of mechanical lifts and everything that goes with skiing.

Montana, in the Valaisan Alps provides a case study of the ski boom. It was a small resort before the war, with a handful of hotels and chalets. It grew vertiginously, and linked up with the neighbouring resort of Crans—which didn't exist till 1928. It played ; host to the World ski races ("The White Circus") in 1987: two new telecabines were installed, new underground parking lots were built, 2000 people worked on organising the races, over a thousand press and TV reporters flooded in. The resort now has 4000 beds in chalets and apartments (yes, including high rise ones), 40 mechanical ski lifts and 190 km of ski trails.

All Alpine Clubs now have an active policy of nature protection; dozens of nature clubs join forces to fight particular projects and occasionally win. The Swiss

Alpine Club has recently appointed a permanent official, solely for the protection of the environment. Shouldn't these be financed by a small head tax on all visitors who would be more than happy to pay. This is not a plea to stop tourism. Than would be impossible anyway. It is a plea to regulate its future development far more strictly and that is not an easy task.

If there is a lesson to be learnt from the experience of the Alps, it is this: protection of the mountain environment needs; determined action and hard work at all levels, from the highest government authorities to the lowliest individual. And the interests of local people have to be well catered for so that they gave a vested interest in protecting the environment. The Swiss, for example, have active programme of aid to mountain communities.

Protection needs some down-to-earth action. Perhaps one way to start is to talk less about the romantic attractions of mountain wilderness areas and more about toilets, garbage disposal and fuel.

Once there is a heavier investment for tourism in the Himalayas, it will be difficult to fight destruction because money talks, and loudly too. Before this happens, it is suggested that the Himalayas should be divided—with the co-operation of all the Himalayan countries into three zones.

The first would be open to intense tourist development; climbers and trekkers would be welcome and infrastructure provided for them. Come one, come all.

The second zone would be open only by special permission to a limited number of climbers and scientists. There are already a number of national parks: these could be upgraded and perhaps even linked?

The third zone would be invilated and kept free of human intrusion. It would be a grade or two above the national park status. This raises some basic questions. Do we want to preserve some corners of the earth free from man's interference or do we want to open up every bit of it to human exploitation? Do we have the right to use up all the earth? Or are we willing to leave a small pocket or two for the snow leopards, ghurrals and the abominable snowmen?

As Reinold Messner wrote recently, "Man, through the centuries, has always chosen places which seemed special and declared them holy and untouchable. There, the gods lived. Today more than ever we have a need of unexplored wildernesses." And our Gods have always lived in the Himalayas. The Crumbling Himalayas.

The Himalayan range stretching from Pakistan to Arunachal Pradesh in India must be one of the most spectacular and dramatic areas on earth. Not only does it embrace the highest mountains in the world, but also mighty glaciers and ferocious rivers, huge foothills, and fertile valleys, extensive deserts and dense forests.

And it is not only the topography that is so impressive, there is a considerable variety of mountain people with different cultures, religions and language. Four decades ago tourism, as we know it today, was almost non-existent. And possibly that has been the major change in the Himalayan activity—the development of number and quality of airfields in the mountains. Nowadays tens of thousands of people are walking the hill tracks, floating down the rivers, driving over the great Himalayan passes, landing in tiny airfields in remote valleys, visiting the sacred places in the mountains.

The impact of all these people visiting the distant

mountain areas has been tremendous. Some of the effects are good and some not; so good. The standard of living of the mountain people has increased considerably from tourist wealth. But the beautiful tracks and valleys have been littered with rubbish and the forests have been destroyed to provide fires for meals and comfort for tourists. Hundreds of tiny hotels do the landscape but toilets are rare and completely inadequate. The Himalayan environment has suffered severly, partly from tourism but more importantly from the substantial growth in the local population. Arable land is increasingly rare and the forests are cleaned and the slopes terraced for the planting of crops. Population growth is a massive problem to overcome—nobody really knows how to deal with it. It is only in affluent countries that the population growth is stable. In the Third World a large family is the only method of social security and there is a constant and desperate race between industries.

As the tourist numbers grow, the impact on the most popular areas becomes more and more severe. But in the huge stretch of the Himalayas there are a multitude of beautiful areas and many of these are completely neglected. Energetic efforts should be made to attract people to the less popular places, to spread tourism thinner, so that their negative environmental impact is reduced. Tourism agencies must be pressurised into ensuring that their customers leave the hills and valleys in as pristine a condition as they were when they found them. One of the great joys of adventure tourism is getting away from the everyday whirl of living, appreciating the peace and contentment of areas, free from public pressure, enjoying the bubbling streams, quiet forest paths, and lonely mountain spaces. We all need to escape from the concrete jungle and return refreshed to the city environment and dream of sparkling waters and icy mountain tops.

The Himalayas are good for all of us but we must not let them be abused. Forests must be preserved, campsites must be kept clean and wholesome, and rivers protected from pollution. We must prevent the wanton destruction of all that is fresh and beautiful and enjoy to the full, the superb beauty of nature in the remote mountain areas.

Environmental Hazards in the Himalayas

The environmental hazards in the Himalayas are many, complex, and interrelated e.g. landslides, droughts, floods, decline in the forest cover, decline in agricultural yield, soil loss, problem of siltation in reservoirs and rivers, extinction of rare species of wild animals, etc. The natural springs are rapidly dying, domestic animals are less productive or unproductive, the fodder resources are limited and natural forests are fast disappearing in the Himalayas.

Blasting of the rocks for the construction of roads, tunnels, dams, colonies. etc and haphazard method of mining have greatly affected the ecological balance. Mining operations are old and traditional and have erred the lush green vegetation cover. The mining operations have also eroded the lush green hill slopes, destroyed leaf mould—a valuable organic fertiliser, polluted streams, and decreased the agricultural production by 60%. In the doon Himalayas, the limestone dust particles from the quarries and lime factories have been estimated at 360 micrograms per cubic meter as against the safety level of 100. Hazardous air and noise pollution have caused many plants to wither away; the forest undergrowth has been cleared, the sub-surface water has decreased, and the mountain side resembles a gigantic skeleton due to blasting.

In higher altitudes, frost edging is an environmental hazard. The glaciers do a lot of abrading and scratching and sometime snow streams and avalanches cause a great loss to life and ecology. In areas of moderate relief of the Himalaya, water erosion is a serious problem as it makes the rock naked; consequently, instability of the hill slopes, silting of the river valleys and floods are very common features. Torrential rains cause mass wasting and huge landslides and landslips. Landslides are very common, often blocking the vehicular traffic and sometimes damaging engineering projects. In the low relief alluvial plains, flash floods, soil erosion and sheet erosion are environmental hazards.

The construction of unimaginatively planned and cruelly executed roads is the biggest factor in destroying the ecological balance. Construction of dams on hill rivers to meet the ever increasing demand of irrigation water and power of the plains is yet another environmental hazard. These projects cause destruction of hills, flood the fertile flatland of the valleys, and create the problems of rehabilitation of the oustees.

The Higher Himalaya

The Himalayas are the loftiest and greatest of all the mountain system in the world. They have from time immemorial held an indescribable charm, inspiration and spiritual significance for the millions living in and around these mountains.

It is well impossible to describe in words the mysticism and the religious aura that surround these mountains. Great Sages and poets have tried to describe the sheer majesty and charm of the Himalayas but could never fully capture in words their majestic; grandeur.

The Himalayan land and water are priceless national assets. The approach of the modern man to explore and exploit its resources is very harsh. If the Himalayan region to meet the increasing demand for fuel, fodder, timber, etc., would be disturbed and it would create environmental hazards in the Himalayas. The question of conservation, regeneration and development cannot be left to chance and the vagaries of political and administrative set-up at the lowest level. Conservation and development should be made mandatory without prejudice to the rights which inhere in the land. In this way, objections or dilatory tactics can be given short shrift.

Geography of the Himalaya

The Himalayas are an awesome chain of mountains that extend approximately to 2500 km. From north-west to south-east and the width is between 250to 300 km. The Himalayan system consists of three longitudinal belts separated by faults, their geological and topographical characters are strikingly different. Immediately north of the plains is the Siwalik Zone. The width of this zone varies from 10 to 50 km. and altitude rarely exceeds 900m. This zone is covered with damp and unhealthy forest. The next zone is known as the middle Himalaya with width ranging from 65 to 75km. and the average height being 300m.

This zone is profusely dissected by the major rivers passing across it. The third belt is the Great Himalaya rising to an average height of 5,200m. The great Himalayan peaks are situated in this zone. There are more than 90 peaks having an altitude of more than 8,000m. This zone has two sub-zones, viz, an alpine zone (4800m.) and an upper snow-bound zone (7500m).

The transverse sub-divisions of the Himalaya are as under :

(i) The Punjab Himalaya—Sutlej to Indus, 590 km.

(ii) The Kumaon Himalaya—Kali to Sutlej, 350km.

(iii) The Nepal Himalaya—Tista to Kali, 800km.

(iv) The Assam Himalaya—Brahmaputra to Tista, 750km.

Dr. Valdiya, an eminent geologist has strongly supported the rise of the Himalaya out of a vast tethys geosyncline that existed where now we see the Himalayas. The soft sediments that lay at the bottom of the geosyncline were pressed, upheaved in several phases. The present configuration and complexity of the structure of the Himalayas is due to the phenomenon of "plate tectonics" and other related orogenic activities. The Himalayas are very young mountains and constitute less than 1/5th of the total geographical area of India. They control the climate and through the rivers the prosperity of more than one third of the total population living in the plains of the Indus, the Ganges and the Brahmaputra.

Water Resources

The Himalayas as the name implies are the "abode of snow". The total ice cover of the Himalayas is around; 16,000sq.km. and this amounts to 17% of their area. A very large number of glaciers exist at present above the snow line, most of which are in the Himalayan range. Glaciers provide water to major perennial rivers of the Himalayan system. Glaciers are formed from the accumulated snow which has survived melting; for a glacier to exist, the snowfall should exceed the quantity that melts in a summer.

There are around 15,000 glaciers in the Himalayas. They vary in length and majority of them are one to five kilometers long. The Gangetic glacier in the Central Himalaya is the source of the Ganges. Its ice surface occupies little over 20 cu.km. of ice. Satopath is another glacier which forms the secondary source of the Ganges. The Brahmaputra which ranks among the mighty rivers of the sun-continent of India, rises in the Great Glacier in the Kailash range in the northern most chain of the Himalaya just south of Lake Tasho in the Tibetan region.

The Indus, the Ganges and the Brahmaputra are the three major river systems having their source in the Himalayas. They drain the country and provide a large potential for the irrigation and domestic use and also for the generation of hydroelectric power. In implementing the development programmes it is necessary to ensure that the adverse environmental impacts do not arise. The total catchment area of these three rivers is about 8,00,000sq.km. Out of the total utilizable surface water resources of 666 milliard cu.m. Over 245 milliard cu.m.is accounted by these three mighty Himalayan rivers.

Indirectly, the Himalayas play a vital role in replenishing the water resources of India. They act as a barrier for moisture laden monsoons and enable to precipitate their moisture on the southern slopes. The melting of the snow and enormous rain keep the drain age system of the Himalayas healthy and further enrich the subsurface flow of water. The Himalayan glaciers are the banks of water. Harnessing of the Himalayan water resources should be taken up in such a manner that ecosystem is not disturbed. The productivity of water resources can be enhanced by trout culture in the free floating water in higher altitude areas and by raising mahasir and mirror carp in the standing waters with extensive use of cage culture technology.

Agricultural Systems

Based on practice, there are two distinct agricultural systems in the Himalayas, i.e., "shifting agriculture" and "sedentary agriculture". Shifting agriculture is being done in the eastern Himalayas while sedentary agriculture is very popular in the western part of the Himalayas. Out of the total area (under cultivation), about 80% is under seasonal crops. The area under tree crop is around 12%. The area under tree crops is shrinking, giving way to season bound crops. Land under the crop in the higher Himalayas has shown an increase of about 15%. Double cropped area in the Himalayas was about 33% on an average, varying from 41% in the Siwaliks to 20% in the middle Himalayas. Sector wise, it is 38% in the central Himalaya and 10% in the Western and the Eastern Sectors.

Cultivated land holdings in the Himalayas are smaller as compared to the holdings in the plains. In the outer Himalayas there are two distinct cropping seasons, i.e., Rabi and Kharif. In the Rabi crops are included wheat and barley which are planted in autumn and harvested in early summer. In the Kharif crops are included paddy, maize, etc. which are planted in spring and harvested in late autumn.

Agriculture in the Western and Central Himalayas require hard work for at least ten months in a year. In the Eastern Himalayas, it is an year round affair. Among the crops, rice cultivation starts from lower altitudes and extends upto an altitude of 1800m. Altitudinally rice is followed by maize and wheat which are grown in slopes above the rice fields. At still higher levels, millets and barley take over.

The Himalayas are the major supliers of temperate fruits, which do well in between 1200 to 2000m. However, the gross area cropped under fruits is very small. Agriculture in inaccessible regions of the Himalayas is difficult and requires considerable hard work. Terracing cultivation is common, even in the steepest slopes. On an average terrace is 3 to 4m long, 2 to 2m wide and 2 to 3m high.

Shifting cultivation or Jhumming practised in the north east has very deep roots and very wide ramifications as the tribal groups system of cultural and ethnic types is typical of the area and has resited change for a very long time. The practice is a low energy budget, a low investment system of subsistence husbandry which uses the comparatively rapid mineralisation and recycling of nutrients provided there are sufficient areas available. Now that there is not enough area for the system to function, degradation is rampant. About 0.542 million hectare area is under shifting cultivation and about 2.64 million tribal population is affected by it. It is now accepted that this practice has to be rationalised.

The limited arable land should be utilised scientifically for valuable cash crops which offer a regional and seasonal advantage and for the development of seed industry for temperate vegetables which cannot be produced in hill zones and adjacent plains. Different cash crops need to be determined based on agroclimated suitability. Horticulture, for which even slopy marginal lands can be exploited, hold much greater promise than food grains. Instead of encouraging only one product (apple), a mix high value low bulk non-perishable fruit cultivation need to be encouraged and popularised in the Himalayas. The traditional agricultural practices are generally faulty; these should be discouraged systematically and ecologically viable scientific techniques adopted. The low products agrarian land should be brought under agro-forestry, farm forestry and agri-horticulture.

Regarding forestry, the forests must be developed beyond the traditional forestry and a significant amount of total demands in terms of fuel, fodder, timber and other tree products can be met from the wastelands and crop fields, if integrated programmes such as social or farm forestry can be implemented at community level considering the needs of inhabitants and site characteristics.

The existing forest land should not be alienated for other land use, particularly agriculture and horticulture. The degraded forest should be enriched by the species already existing in such biotopes as these are well acclimatised to the locational factors compared to the exotic species. Free range grazing as practised in the region is injurious for regeneration of forest. Local inhabitants should be educated about the merits of stall feeding and rotational grazing. Afforestation programme should be encouraged; and popularised as major component of the rehabilitation and conservation strategy. There should be adequate stock of seedlings and seeds to maintain an ecological balance, it is necessary to build up productive assets and to increase the supply of consumer goods, fuel, food, fibre, fodder, and products of resin, etc.

Animal Husbandry

The life and environment in Himalaya is a symbiotic relationship between agriculture, forestry and animal husbandry. The Jersey cow has been found as the best milch animal for Himachal Pradesh while buffalo for Central Himalaya. The local livestock can be improved; by cross breeding the unproductive animals should be replaced. The Angora rabbits may be introduced as an alternative to local sheep for wool to particular extent. The bovine population may be reduced by castration of local bulls. The existing mode of open feeding should be stopped. The poultries to be encouraged as considerable income is generated through these.

Grassland

Silvipastoral or Silvo-hortpastoral systems in the Himalayas are more useful and profitable compared to grass development alone. Effective protection of grazing land would increase the herbage cover and herbage yield. Fertilisation and introduction of grass legume mixtures in mid hills and cocksfoot, timothy, personal rye grass and contucky grass in sub-Alpine and Alpine grasslands can regenerate the degraded grass resources. Introduction of plantation of fodder trees on terraces and private lands shall improve the position of green top feed during the lean periods. Thus, planting improved grass varieties or strains coupled; with more efficient management of grasslands is vital in meeting current and future needs of herbage providing protective soil cover.

Tourism

The Himalayas offer tourists a variety of scenic beauty. To them it is a paradise, its gushing rivers are laden with trout and the varieties of fish. To trekkers and hikers it is a challenge, its canyoins, deep valleys and escarpments are eye-catcher. The snow-peaked mountains, gliding glaciers, grass meadows and hot springs invite tourists from remote corners of the world. The ever increasing tourist influx is creating great impact on ecology and environment. The problem of litter, noise pollution and damage to natural assets is causing great anxiety to environmentalists. The unmanaged influx of tourists destroys the natural architecture of the rocks and natural surroundings. The development and construction of new hotels involves the destruction of vegetation, felling of trees and erosion of the environment. Tourism should be considered as on alloy and not an adversary in conserving the ecology in the Himalayas. It should be in the fitness of the things if master plans are prepared and new areas are develped to meet the unending demand of tourists resorts. There should be a close coordination among the planners and the promoters of ecology and environment of the Himalayas.

Land Resources and Tourism

The construction of roads and other infrastructure are indispensable for efficient transport and communication system for tourism development, but these unavoidable activities are mainly responsible for destruction of Himalayan land resource along its vegetation and animal wealth. A careful study shows that the construction of 44,000 km. long roads in Himalayan region produced 2,650 million cubic metres of debris and each kilometer of already constructed roads generates; 550 cubic metres of debris by landslides and rockfalls. Thus nearly 24 million cubic metres of debris slide down damaging vegetation and chocking the springs etc. (Valdiya 1985). These activities accelerate hill slope instability deforestation, soil erosion, pollution of water and air, etc, especially along the road sides and around the most of the Himalayan tourist resort like Nainital, Simla, and Mussoorie etc.

The use of powerful explosives like dynamite and frequent heavy traffic create vibrations which casue cracks on rocks and earth which in turn lead to land slides and erosion along with adverse impact of the flora and fauna. The construction of roads etc. along geologically active zones like at Kaliasur, Satpuli (Garhwal) is also responsible for degration of land. The construction for accomodation facilities like rest houses, tourist bunglows, buildings and seasonal accomodation units, recreation centres,etc., have posed the most serious menace to the land resources of the region; not only does it accelerate soil erosion but it also causes damage to crops, animals and plants, housing and public properties. It is also a contributory factor in causing floods with heavy siltation in valleys and adjacent areas. In Chamoli district alone from 1971 to 1978, 48 human lives, 949 heads of cattle, 601 houses, costing nearly about 173 lakhs of rupees, as also the crops, were destroyed by mass movements. In the year 1972 owing to the landslides, the Gohna Lake was formed and its breaching has resulted in a large scale destruction in the Alaknanda valley and the plains of U.P. The flash flood engulfed 2000 sq.k.m.

paddy crop, 10k.m. roads, 6 motor bridges and 24 buses, 366 houses while 138 villages were effected; by landslides alone in the year 1979.

Agriculture and Tourism

Agriculture is one of the main occupations of the Himalayan inhabitants, but it does not give quick and heavy economic gains like tourism. At present during the tourist season (mainly May-June) tourism has become the main source of income, as a result of which agricultural system has miserably regressed. Enormous agricultural land is being occupied for development of infrastructure facilities. These activities create dependence system in addition to adverse impacts on economy and ecology of the area. These activities create dependence system in addition to adverse impacts on economy and ecology of the area. The farmers who are living in the vicinity of these tourist resorts leave their traditional farming and either work or conduct different types of business for prompt and easy economic benefits. In many such developing and well developed tourist places, tourism has almost overlapped agriculture. Fields around the famous tourist places like Mussorie, Nainital, Simla, Darjeeling, etc.are the result of this type of tourism impact.

Forest Wealth and Tourism

The beauty of the Himalayan forest, alpine meadows and peaks is the main sources of tourist attraction. The Continual flow of tourists, mountaineers, trekkers, etc. is likely to have positive as well as negative impacts on these natural resources to a certain extent. Mythological beliefs and tradition which are prevalent in the Himalayan region are also responsible for adverse effect on the flora, specially in those localities adjoining the shrines. Religious tourists habitually collect flowers and colurful herbs to offer to temples like Gangotri, Yamnotri,

Badrinath, and Kedarnath, etc. Consequently, many species which are part of our cultural and religious heritage are beginning to get wiped out. It was noticed by the authors during the Roopkund trekking in Chamoli district that at the Bedney Bugayal and Roopkund (places of mythological importance along the trek) thousands of Brahmakamal flowers (Saussurea obvollata) were offered which were plucked from vicinity area. A recent sample study of Surkanda Devi (a place about 25km. from Mussorie towards Tehri), which is situated at a most picturesque spot in the highest (2750m) peak of outer Himalayan range, indicates that about 10,000 people gather there for a week on the occasion of Dashera (celebrated festival of Hindu) and practically every one offers few flowers with twigs of Ronsli (Abies Pindrow) to Goddess Surkanda along with their other gifts. Every visitor roughly plucks ten flowers and two twigs. In this way about 10,000 flowers and 20,000 twigs are lost annually, as a result of which a virgin Abies pindrow forest and densely grown different colourful flowers which we have seen during childhood (about 23 years back) have nearly disappered now, and in their place scrub forest and flowers like Primula denticulata, Anemone vitifolia, etc. are sparsely found. In addition to this, severe trampling by visitors is also reponsible for damage.

A recent field survey of the world famous valley of flowers also shows that its flora and fauna was under major attack and influenced adversely after 1975 when this area was opened to visitors. Due to increasing rate of tourism traffic and direct (like trampling and plucking etc.) and indirect activities of tourists, many important and charming species of flowers like blue-scented Primulas (most beautiful), red Potentilla spp., creamy Anemone spp., blue Meconopsis spp., etc. are now found in reduced abundance along the treks, road sides and other easily accessible areas. This degradation will definitely occur in the entire area with increasing number of tourists. In addition to this, other factors like overgrazing, low grazing etc. are also responsible for ecological imbalance. Overgrazing prevents plants from flowering and fruiting willow or ungrazing accelerate the growth of unnecessary weeds and grasses which create hinderence in growth and blooming. The growth of unwanted grass and weed has become a major problem of the valley after the ban imposed on grazing by the government and it seems to be only one reason for low flowers occurence in the valley even in peak season (August-September) in comparison with the situation obtaining some years ago.

This fact leads to the conclusion that a scientific and controlled grass cutting or grazing is necessary for maintaining the importance of the valley of flowers. If the sensitive ecosystem of the valley will not be given proper importance and attentions along with the promotion of tourism, Smythe's silent valley of Garhwal Himalaya may one day cease to show its visual spectrum, of nature's beauty. The damage in the Himalayan environment has reached a critical level and beyond the carrying capacity; therefore, in order to overcome this problem, it is the need of time that certain more landscapes with unique type of vegetation and wildlife should be conserved as "Biosphere Reserves".

The tourist concentration in the Himalayan alpine pastures is increasing day by day, more than the carrying capacity, while the area under vegetation is decreasing due to accelerated human activities.

A considerable area of virgin forest is also being replaced for providing necessary amenities to visitors for tourism promotion. It has been estimated that 234.25 sq.km. forest area has been destroyed only by road construction from 1951 to 1979 in the U.P. hills (forest statistics 1981). Deforestation not only result into resource depletion but also deteriorate fresh and underground water, wildlife and other related resources along with acceleration of disastrous phenomena like heavy soil; erosion and landslides etc., which mainly provoke flash floods.

Wild Life and Tourism

The forest provides the best shelter and source of food for most of the wild animals, hence the quality of forest affected by vast tourism directly influences the wildlife of the respected area. The ever increasing influx of tourists exerts an adverse impact on the unique Himalayan wildlife.

Tourist activities like invasion in wildlife habitat, close view watching and photography, hunting, poaching, and other disturbances etc. have created a problem for wild animals. Mammals and birds are affected in large extent—more than reptiles, insects and other small animals. Owing to the large interference of man in the habitat and life style of animals specially in national parks, wild life sancturies etc., the ferocity of animals is being lost which is the only quality of wildness that differentiate them from domestic animals. The act of providing bate to tiger or lion for close view in national park etc., spoil the quality of wild life and is destruction of wild animals in the name of tourism. These activities further affect the feeding and breeding habits and ultimately the entire life cycle. The small animals die due to heavy and frequent traffic which affect the food chain in wild life system.

The habit of tourists to provide ready made food to wild animals is also responsible for loss of ferocity. During the year 1975 in the Corbett National Park a wild boar popularly known as Buddha (innocent) had become a purely domestic one. Owing to the availability of ready made food like bread, biscuits, chapati, etc. from tourists he had later on developed the habit of stealing edible items like lunch packets, chocolates, jam butter, even tinned meat and fish, sauce and drink bottles, etc., from tourist tents, kitchen, or even open car, jeep etc. thereby a purely herbivourous wild animal had changed his wild habits. Recently in the year 1987 is has been observed by the authors in Corbett National Park that young chittal (spotted deer) is also frequently visiting the Dhikala complex of the park to eat cooked food material like bread, etc., from the hand of tourists and officials.

It is also noticed that during night time; few wild animals like wild boar (Susscrofa), fox (Vulpeseulpes), chittal or spotted deer (Axis) etc., reach close to the Dhikala complex and feed on thrown-away ready made food or garbage etc. sometime even together also. So it is necessary to protect these animal habits and habitat alongwith their natural ferocity to maintain the wildness and ecosystem balance of natural food chain system between predator and the prey or between the herbivorous and herbs for total ecological system. The crowded gathering of people religious places and resorts bring marked change to ecosystem as a whole. The direct and side products of tourists like litter, polythene bags and wrappers, garbage piles and related materials attract rodents and crow, which in turn draw bigger animals and birds, thus changing the entire habit and habitat.

The increasing tourist's craze for the Himalayan wild life and their products is also responsible for its destruction as it increases the demand of stuffed animals, ivory ornaments, furs, tails, quills, feathers, horns, skins, musk and wild-life souvenirs, etc. This demand provokes the curio trade, resulting into poaching and slaughtering of animals. The musk deer, (Moschus moschiferus), blackbear (Selenaretos thivetanus), snow leopard

(Panthera unica), leopard (P. pardus), bharal (Pseudois nayar) monal (Lophophorus impejanus), etc. are few well known animal species which are in great danger due to poaching.

To prevent or reduce degradation of the Himalayan natural resources, there should be scientific and well planned growth of tourism in which experts of different disciplines like geology, botany, geography wildlife, environment, etc., and technocrats like engineers, architect etc. along with local people should have close involment in planning. There should be recreation between natural resources and tourism development which should be according to the carrying capacity of particular resorts, so that space action does not affect their ecosystems and ecological balance. If these protective measures are not given due importance in tourism planning and development, the Himalaya would be damaged beyond the limit of repair of regeneration.

Industrialization

Industrial development in the Himalayas is slow but the rapid population explosion and urbanizations has changed the trend. In the Kashmir Himálaya, the Himachal Himalaya, the Kumaon Himalaya, the Assam Himalaya and in many other parts of the Himalayas fruit processing, wool processing cottage industries, cement industry, paper and pulp industries, resin processing and other forestbased industries, such as watch factories, tool factories, television factories etc., with pollution control devices would be vital for the rapid economic growth of the Himalayan belt.

Measures to Preserve the Ecology

The Himalayas provides a variety of natural resources to the sub-continent of India, including the life-giving water These resources have been threatened largely by human negligence and activities. In order to improve and regenerate degraded hill environment, cientific approach involving hazard zone mapping and banning engineering development activities in such zones, drainage control and dewatering measures, protection of slopes through vegetation, modification slopes where necessary, application of engineering technology and improved agronomic practices should be carried on priority basis.

The ecological balance can be maintained and improved by reforestation, by checking the illegal felling of tree, controlling the course of floods by construction bounds, retaining walls, schecking landslides, rocketries, sinking of the soils, protecting the watersheds, etc.

Laws should be enacted to stop indiscriminate timber trade, installation of new cement factories, stone crushers, state factories, industries, etc, which in turn cause air, water and noise pollution. The mining operations should be done using modern methods and techniques to minimize the pollution of atmosphere, the ruthless and haphazard mining activities greatly affect the health of the workers and also create ecological imbalance by rock debris, dumps and quarries' wastes. The greenery of the soils is lost and fertility of lands is decreased. Constructional and developmental works also destroy the natural setting of the scenery and landscape.

Soil and land use inventories need to be taken up on priority basis and the Himalayan belt may be divided into viable seacoast for the purpose of land use, planning and management. Modern techniques of remote sensing and aerial photo interpretations are very vital ion conserving the soil and soil programme. Soil maps may be employed to carry out the afforestation programmes and proper landuse. Aerial seeding should be carried out to reclaim the barren lands.

To prevent or reduce degradation of the Himalayan natural resources, there should be scientific and well planned growth of tourism in which experts of different disciplines like geology, botany, geography, wildlife, environment, etc and technocrats like engineers, architect etc. along with local people should have close involvement in planning. There should be correlation between natural resources and tourism development which should be according to the carrying capacity of particular resorts, so that the space action does not affect their eco-system and ecological balance.

If these protective measures are not given due importance in tourism planning and development, the prosperous heritage and environment of the Himalaya would be damaged beyond the repair of regeneration.

To protect our environment we have to protect our wild-life, rare birds and unique plants. Proper education on environmental rehabilitation and conservation must be initiated at the grass root level. Rehabilitation and conservation programmes should include such measures as finding alternative sources of energy, self employment, agriculture, sericulture, horticulture, social forestry, cottage industries, small scale industries, etc. The human and animal husbandry programmes should also be taken up to improve the lot of the Indian village folks. Women are the backbone of human ecosystem in the Himalayan hills. It is, therefore, necessary to improve their lot. An all-out effort should be made to maintain the enchanting beauty of the Himalaya.

Environmental Management of The Himalayas

It is only during the last few decades that man has begun to exploit the vast and varied natural resources. The great rivers which originate from the springs and the snowfields possess immense potential for a multiplicity of uses. At several places these rivers have been tapped and plugged to generate hydroelectric power, to construct dams and generate power, to irrigate the land, and to transport the material through the rivers. A proper management of these rivers and other natural resources to benefit the inhabitants of the holy mountains is very vital. The problems of environmental management are diverse and require special attention.

Conserving & Managing the Depleting Resources of Plants

The Himalayas extend from the west to the east for about 2,500 km. and they very in width from 250km. to 300km. These mountains have the world's highest peaks. The diverse landscape soils and climatic sectors, support several types of vegetation. Some of the plants are very characteristic of tropical, sub-tropical, cool temperate, temperate and alpine conditions. Some of the regions also support desert or semi-desert vegetation. The Himalayan ranges can be divided into three major botanical regions i.e. the western, the central and the eastern Himalayas. Of these, the western region is further subdivided into the western Himalayas, Himachal Pradesh and the north western Himalaya. The central Himalayas more or less, conform to the mountainous region in Nepal. The eastern Himalayas includes Sikkim, Bhutan, Arunachal Pradesh, etc.

Considering that the Himalayan region is richly endowed with a wide range of plants and is, in fact, one of the centers of origin of certain important agricultural and economic plants, an earnest need is to survey plant wealth, especially rich in genetic diversity. The survey, however, should not be merely a taxonomic one but a dynamic one carried out by multidisciplinary experts with the object of conserving the rare species and those species which are vital in the field of medicine, forestry, agriculture. etc. An effort should also be made to check them growing wild.

The exotic plants, e.g. rhododendron, wild roses, coniferous trees, and rare herbs and food crops should be genetically upgraded to meet the requirement of food.

The temperate Himalayan belt offers great potentials for forage grasses and legumes, as such the natural grasslands of low productivity can be replaced with highly productive by farming systems. These untapped grassland resources of the Himalayan region can perhaps make the largest contribution to its economic development. As the prosperity of a nation is chiefly determined by its capacity to utilize and conserve its resources, it is therefore, necessary to undertake the steps based on the physical and biological factors of the region so as to formulate a plan for the rapid development, economic progress and sound management of the Himalayan plants.

Conservation & Management of the Wild Life

The wild mammalian fauna of the Himalayas shows a tridimensional zonation, i.e. (a) Longitudinal, (b) Latitudinal and (c) Altitudinal.

In each major zone, number of subzones can be recognized. They show interesting adaptations and extralimital affinities. The Himalayan mammals are provided with long and thick hairy covering atleast in winter to cope with the severe cold. Several animals resort to migration to lower heights during winter. Animals confined to the coniferouns regions show a darker colouration as on adaptation to increased humidity and also as a protection against ultraviolet radiation, but the animals capable of living above the forest zone tend to be paler: e.g. the snow leopard, the Horsfield's leopard cat, the Trevelyan's leopard cat, the pillar's cat, the snow bear, the Himalayan ermine and pikas.

The member of Artiodactyla which have speed as the chief means of escape from the enemy are well adapted for quick movements on the steep and rocky slopes of the mountains by means of their splayed and pointed hooves. Some carnivores like tiger, lion, panther, cat, etc., show highly interesting adaptations of stalking the prey on the difficult terrain of the high altitudes. In the case of the snow leopard, the ears are very small and the eyes are placed high up on the head so that the animal is able to reconnoitre for prey and locate it with a minimum possible exposure of head. In the case of the pillar cat, the short ears are set very wide apart and low down on the sides of the head so that their inner and upper borders hardly rise above the forehead which is highly flattened. The adaptation is of vital advantage for stalking the prey without the exposure of the predator from behind the rocks.

The wildlife in the Himalayas has been greatly affected by rapid development and urbanization. The rapid deforestation has also resulted in the extinction of wildlife. The wholesale degradation of the Himalayas, indiscriminate poaching of wildlife for commercial, medicinal, and industrial purpose has greatly reduced the stock of the wildlife. The illegal trade in protected animals now has higher profits than drug trafficking. A kilo of rhino horn costs about Rs. 2,50,000/- and a kilogram of musk can fetch upto Rs.7 lakhs in the international market. The number of animals killed in laboratories for academic and research work is difficult to ascertain. Cosmetics, shampoos, etc; from big industrial concerns are first tested on the eyes of rabbits resulting in excruciating pain. The trade in hides, skins and live animals is indeed cruelsome. The use of pesticides and insecticides to eradicate deleterious insects cause harmful impact on the wildlife of the Himalayas.

For an effective and successful conservation of wildlife, it is necessary that deforestations, spraying of pesticides, insecticides, trade in animals and experimentation on animals should be legally stopped. The number of parks and sanctuaries should be increased to preseve the vanishing rare species. The proper management of the sanctuaries and national parks is very vital. They should be protected from poachers and those who trade in animals. Factors limiting their population should be tackled seriously.

Among the pests are fruit bats, rodents, lagomorphs, etc. Their control, conservation and proper management can be done by the agriculturist and the silviculturist. The population of the insectivorous bats can be increased by providing roosting boxes for these highly beneficial animals. A well planned management programme is bound to conserve the fauna of the Himalayas.

Livestock Resources and their Management

The Himalayan region is an intricate ecological complex stretched over an area of 61 million hectares of which an ared of 17.8 million hectare is under forests and another of 0.7 million hectares in covered by alpine pastures. The ecosystem ranges from a high humid thick jungle belt in Assam to the dry sub-zero desert wastelands of Spiti and Ladakh. The entire Himalayan belt can be sub-divided, on the basis of bioclimatology, into three belts i.e. the dry cold deserts of Ladakh and Spiti, the true temperate hills, and the humid and the high rainfall hills.

Consistent with this ecological pattern, the livestock resources have developed, having become both dependent on man and independent of him. The Himalayan belt supports 20 million cattle, 10 million buffaloes, 2.9 million sheep, 5.94 million goats, 0.12 million horses, donkeys, etc; 1.27 million pigs. 17.38

million poultry and other birds and other livestock comprising 1.18 million heads which include yaks, charoes. etc. The animal resources are large and can be a source of proteins for human consumption, a source of fibre, leather, dairy products. meat products. etc. A sound environment requires gradual replacement of animal resources by better stock and through the addition of new highly productive genes. A crop-livestock ecosystem mix which can maximize the yields of animal products without disturbing the environment should be adopted.

Agriculture, Soil Conservation and 'Landuse

The Himalayas act as a barrier for the monsoon and cause precipitation to enable millions of people to carry out agricultural operations in the extr-peninsula and Indo-Gangetic plain. The annual rainfall varies in different ecological frames and structures from 5 cm in Ladakh to 400 cm in Arunachal Pradesh. Precipitation in the form of snow is very common in the higher atltitudes. The soils are mostly acidic at higher altitudes in the eastern Himalaya. They are found in association with nonacidic soils at mid altitudes of the western Himalaya and at low altitudes of U.P. Himalaya. The meadow soild are also acidic in character.

An acute problem with the soils is their erosion. The degree or intensity of erosion is very high and is of the nature of sheet erosion, gully erosion, etc. Owing to steep slopes, erratic landuse, indiscriminate grazing, lopping and felling of trees at random, flash floods, silting etc. the problem has become acute. Improper landuse, small and scattered holdings, limited irrigation and old (or conventional) methods of agriculture need due attention. A more rational land use of the resources can be introduced in the region by carrying out systematic soil surveys and preparing soil and land use inventories. The demarcating lands suitable for agriculture or horticulture or both. The chopping system and crop rotation studies do the least damage to the environment while bring to the farmer maximum returns from his land. Experiments should be initiated to determine the ways and means of aligning the rows of plants and the shape, heights and contours of the terraces to minimize losses of the precious "top soil" due to runoff. etc. Considering that domestic animals are indispensable to the people of Himalaya, a programme should be launched to develop pastures and suitable leguminous plants, and fodder grasses should be introduced in the region. Keeping in view the spiral rise in population, it is recommended to supplement the food production in the region by encouraging the fish industry and fish-culture. The trout fish is becoming very popular, the economic lot of the people living in the vicinity of the lakes, springs, ponds, etc. where the trout fish is found. There is also need for overcoming Jhuming way of cultivation in the eastern Himalaya. The proposed management and soil conservation techniques would certainly make the region agriculturally prosperous and self-sustaining.

According to an eminent scholar, Dr. M.S.Swaminathan, the regions of the Himalayas should be studied with reference to weed plants, their biology, their relationship with the crop plants, their distribution, control, diseases, etc. There are many medicinal and aromatic plants which need special investigation for making economic use of them. Studies on orchids, several type of bulbous plants and mushrooms will definitely help to improve the economy of the people in these regions. Studies on the harvesting, storing and sale of crops is of utmost importance. The problem of rats, bamboo rats and other rodents should also be solved. The fesibility of alternative land use and constraints in the adoption of new techniques should be given priority. It is an opportune time to fight the unholy tripple alliance of pests, pathogens and weeds.

Managing the Depleting Forest Resources

The place of forests in the Himalayan environment depends upon the vegetational and geological evolutionary history, longitude, latitude, altitude, drainage patterns, climatic patterns, and soild conditions. Their role in regulating precipitation, streamflow, subsurface flow, supplying food, fodder, fruits, etc. is vital. Their preservation and maintenance is of paramount importance to maintain the ecological balance. In order to conserve the forest resources and to check their rapid depletion, it is essential to link up forest reources with forest-based industries on the basis of physiographic, industrial and economic catchment concepts. The rural economy can be improved by several schemes of social forestry. The wastelands and degraded lands be employed for "organized forestry," "multipurpose farm forests," etc. Keeping in view the importance of vegetation cover and sound landuse practices for soil and water conservation, it is felt that there should be a close co-ordination among the irrigation, soil conservation and forest departments both at the State level and at the Centre. This shall help to ensure optimum social and economic returns to the local inhabitants without bringing imbalance in the ecosystem. The evolving of sound management practices should form the core of all scheme pertaining to afforestation and regeneration. Intensive study should be made to identify the suitable species of trees for wastelands and the degraded lands.

Water Management

The problem of silt discharge is very acute in the Himalayan streams. An adequate network of gauging stations be established in the hill streams. These stations should be equipped with adequate instruments, equipments, etc., to monitor the qualitative and quantitative analysis of silt discharge and sedimentation throughout the year. An effort should also be made to control siltation in all catchment areas.

Recognizing that the utilization of the irrigation potential of the Himalayan region is sub-optional, it is imperative to survey the irrigation potential for sound water management. Since srings are vital sources of drinking water in large parts of the Himalayas, it is, therefore, very necessary to undertake the hydrological investigations to ensure a regular utilization of the spring water. To prevent water run off and judicious use for protective irrigation, small reservoirs to suit local conditions, soil, climate, topography, etc. should be constructed. As most of us know the Himalayan glaciers are the source of several streams, it is, therefore, necessary to make an exhaustive study of the Himalayan glaciers and also prepare an inventory of glaciers, glacial debris, glacial moraines, etc.

Managing the Physiographic and Catastrophic Hazards

The geologically young are of the Himalayas, the presence of residual stresses and highly deformed and tectonized condition of the rocks make them highly susceptible to weathering, erosion and damage by structural and seismic activity. The developmental activities in the Himalayas have to face the challenge of these geological hazards, and in some cases the developmental activities, such as the impounding of large quantities of water in reservoirs, tunnelling operations and the addition of silt load of rivers by road cutting, lead to the intensification of the natural dynamic processes. A sound management in the direction requires proper planning and designing of the engineering projects, their environmental considerations, protection or rehabilitation.

Subaerial denudation is another physiographic hazard in the Himalayas. The weathering is mostly due to block disintegration and thrawing. This causes the production of vast quantity of Scree or talus. The debris or scree slips down along the slopes under the action of gravity. The debris falls on the fertile fields, stops the free circulation of water, roads are blocked and landslides of intense magnitude are caused. The Badrinath area of Uttar Pradesh presents a typical example of subaerial denudation and landslides.

Intense grazing has also accelarated the processes of soil erosion and this has made the slopes barren, naked and prone to sheet erosion and gully erosion. Enormous quantities of silt and sediments fill the river valleys and result in flash floods. The catastrophe of the river Brahmputra alone annually carries a suspended load of 800 million tonnes and the average sediment yield from its catchment in India is of the order of 26,000 hectares metres.

The Siwaliks in the outer Himalaya experience mass wasting and this results in huge landslides. Such landslides are very frequent near Nashri Nala (Kud-Batote) in Jammu, near Poonch (Jammu), Malori on the Batote-Bhadarwah road of Jammu, Mussorie highway on the Chakkir-Dalhouse road and on the hilly regions of the eastern Himalaya. The landslides have also been observed in the Kumaon and Garhwal Himalayas and they cause a great loss of life and property.

Dr. K.S.Valdiya, an eminent scholar of the country, is of the opinion that the tectonic stresses which heaved up the mighty Himalayan mountains have not died down as yet, and provide an additional impetus and vigour for the dynamic processes. These processes are manifested by intense seismic activity, mass wastage, floods, transgressions, sinking, subsidence, regressions, etc. Most of the earth-quakes that have been felt in the extra peninsular region are due to the dynamic forces and till date million of people have died besides an irreparable loss to landscape and property. The earth temors also cause huge landslides. The creep movement in the areas of Chamoli district of Uttar Pradesh in early 1968 resulted in a 40 metre high blockade across the Rishiganga river near the village of Reni(U.P). The lake thus formed was silted up by May, 1970 and the blockade was breached during July, 1970, and the floods in the Rishiganga river created environmental hazard. About $0.175 km^2$was reactivated owing to the erosion of the debris derived from a prehistoric slide by the Karamnasa Nadi—a tributary of the Alaknanda, resulting in a debris avalanche, which washed away the motor road built above and blocked a part of the course of the Alankanada river.

Major slides occured also in other parts of the Himalayas causing blockades, floods, disruption of roads, etc. Avalanches are also very common in the higher altitudes of Leh, Kargil, Gurez etc. in Kashmir, Spiti in H.P., in the higher reaches of Kumaon-Garhwal and Badrinath area of U.P. These cause disruptions of telecommunication lines, electric poles, destruction of engineering structures, disruption of roads by the accumulation of snow and ice. A proper appreciation of these dynamic processes and their scientific evaluation in time can decrease the intensity of loss, if not stop it altogether.

The rock waste derived from the construction and excavations of the roads, bridges, reservoirs, tunnels, etc. should be placed at an area which may neither affect the cultivated lands nor the flow of a stream. The appropriate use of such a waste is in the construction of rock-filled and concrete dams or in road dressing. The environmental management of the Himalayan landscape is fraught with complex interrelationships between natural entities and the evolutionary processes, and between human activities, government concerns and programmes of resource management. The human interference for developmental activities in the Himalayas has to be in consonance with the environmental conditions. As the Himalayan terrain has some of the best tourist resorts, e.g. Kashmir, Nainital, Dalhousie, Mussori, Darjeeling, etc., it is essential that the developmental activities should be planned in such a way that their aesthetic appeal is not lost.

The maintenance and development of roads which play a significant role in the ecology of the hill terrains, should be given due attention and Central and State teams should work in close co-ordination for the development of roads.

It is finally recommended that long term basic research into the functioning and dynamics of the Himalayan ecosystem should be initiated for an everlasting and sound environmental management of the the Himalayas. The following priorities for research are identified :

(i) The problems associated with the current massive erosion and landslides occuring in the hilly terrains of Himalaya.

(ii) The problems associated with the explosion in population, rapid urbanization and migration of masses in the mountains regions.

(iii) The problems associated with the rapid changing faunal and floral stocks in the Himalayas.

(iv) The problems associated with tourism and the adverse effects on the environment.

(v) The problems associated with ecological and environmental awareness and establishment of "Biosphere Reserves".

The intense investigation of the above referred problems would help in achieving the destined objects of an excellent and sound management of environmental problems in the Himalayas.

6 Atmosphere

The atmosphere refers to a protective blanket which is able to nurture life on the earth and protect it from the hostile environment in outer space. It constitutes the source of carbon dioxide for plant photosynthesis and of oxygen for respiration. It provides the nitrogen which nitrogen-fixing bacteria and man's ammonia manufacturing plants employ to produce chemically bound nitrogen essential for life. As a basic component of the hydrologic cycle the atmosphere transports water from the oceans to land, thereby acting as the condenser in a vast solar-powered still. Unfortunately, the atmosphere also has been a dumping ground for many pollutant materials—ranging from sulfur dioxide to aerosol can Freon—a practice which causes damage to vegetation and materials, shortens human life, and possibly alters the characteristics of the atmosphere itself.

The atmosphere forms an insulating blanket around the earth. Without it the temperature at the equator would rise to 180°F during the day and drop as low as—220°F at night. It burns up meteors that would bombard the surface of the earth from space. Without the atmosphere there would be no sound and no flight. There would be no conventional long-distance radio communication, for this is dependent on the electrons in the upper atmosphere. Without air there would be no lightning, no clouds, no wind, no rain, no snow, and no fire. The surface of the earth would be as bleak and sterile as the moon.

The atmosphere serves a vital protective function. It absorbs most of the cosmic rays from outer space and protects living things from their effects. It absorbs most of the electro-magnetic radiation from the sun. Only radiation in the wavelength regions of 300 to 2500 nm and 0.01 to 40m gets transmitted to any appreciable extent by the atmosphere. The first region consists of near-ultraviolet, visible, and near-infrared radiation; the second consists of radio waves. One segment is the *optical window,* consisting essentially of the visible spectrum of light, from near-ultraviolet to near-infrared. The other is the *radio window,* consisting of radio waves from about 1 centimetre to 40 metres in length. It is particularly fortunate for life on earth that the atmosphere filters out tissue-damaging ultraviolet radiation below about 300 nm.

The atmosphere has been essential in maintaining the heat balance of the earth. The atmosphere absorbs infrared radiation emitted by the sun. It also absorbs energy re-emitted from earth in the form of infrared radiation. Therefore, it serves as an important heat stabilizing function and prevent the tremendous temperature extremes which occur on other planets and moons lacking substantial atmospheres.

It is impossible to define the limits of the atmosphere because the atmosphere becomes progressively tenuous with increasing distance from the earth. There is no boundary between the atmosphere and the void of outer space. However, 75 per cent of the earth's atmosphere lies within 10 miles of the surface and 99 per cent of the atmosphere lies below an altitude of 30 kilometers (19 miles). The total mass has been estimated at 5500 trillion tons.

Composition of the Atmosphere

The components of the atmosphere may be divided somewhat arbitarily into major, and trace constituents. These components expressed for pollution-free dry air at ground level are given in Table 1.

Table 1

Components in Pollution-Free Dry Air at Ground Level Expressed as per by Volume

Components	*Volume per cent*
Major Components	
Nitrogen	78.08
Oxygen	20.95
Minor Components	
argon	9.34×10^{-1}
carbon dioxide	3.14×10^{-2}
Trace components	
neon	1.818×10^{-3}
helium	5.25×10^{-4}
methane	2×10^{-4}
krypton	1.14×10^{-4}
nitrous oxide	25×10^{-5}
hydrogen	5×10^{-5}
xenon	$0\text{-}8.7 \times 10^{-6}$
sulphur dioxide	$0\text{-}1 \times 10^{-4}$
ozone	$0\text{-}2 \times 10^{-6}$
nitrogen dioxide	$0\text{-}2 \times 10^{-6}$
ammonia	0-trace
carbon monoxide	0-trace
iodine	0-trace

In addition to the components in Table 1, air may be having from 0.1 to 5 per cent water on a volume basis, with a normal range of 1 to 3 per cent. The density of the atmosphere decreases sharply with increasing altitude as a consequence of the gas laws and gravity. Over 99 per cent of the mass of the atmosphere is found within approximately 30 km (1 km = 0.62 miles) of the earth's surface. The total mass of the atmosphere is nearly 5.5 x 10^{15} tons. Large as it is, this enormous mass has been only approximately one millionth of the earth's total mass.

The characteristics of the atmosphere have been found to vary greatly, particularly with altitude. Other factors introducing variability include season, latitude, time and even solar activity. Temperatures in the atmosphere may vary from as low as as 138°C to over 1700°C. Atmospheric pressure drops from 1.00 atm at sea level to 3.0 x 10^{-7} atm at 100 km above sea level. Due to these variations, the chemistry of the atmosphere varies greatly with altitude. In addition to temperature differences, the mean free path of species in the atmosphere (mean distance travelled before collision with another particle) increase over many orders of magnitude with increasing altitude. A particle having a mean free path of approximately 1 x 10^{-6} cm at sea level is having a mean free path which is exceeding 2 x 10^{6} cm at an altitude of 500 km.

Major Regions of the Atmosphere

The science of planetary atmosphere is termed as *aeronomy*. In addition to dealing with the chemical composition and reactions of these atmospheres, it involves their movement and thermal balance. In order to understand atmospheric chemistry it becomes essential first to have some understanding of the physical characteristics of the atmosphere.

It is possible to divide the atmosphere into a number of different regions depending upon the system of classification. The most general classification has been that of the lower atmosphere (up to approximately 50 km) and the upper atmosphere. Until the development of high altitude rockets and earth satellites, only the lower atmosphere was accessible for study. However, use of these devices has caused a quantum jump of man's understanding of atmospheric phenomena in recent years.

The upper region of the atmosphere is having species appreciably different from those found in the lower region. As the lower regions are relatively homogenous in composition, one system of classification divides the atmosphere into regions called the *homosphere* (having little variation in composition) and the *heterosphere* (having a high variation in composition).

The system most commonly used is to divide the atmosphere upon the basis of temperature changes. These temperature regions and their characteristics have been summarized in Table 2 and have been depicted schematically in Fig. 1. In addition to the regions listed, a fifth region, the *exosphere,* is sometimes defined. Some of the species in the exosphere attain sufficient kinetic energy to break loose from the gravitational field of the earth and go into outer space.

The boundaries of the troposphere have been found to be influenced by the number of factors, including temperature and the nature of the underlying terrestrial surface. Even during a one-day period the upper limit of the troposphere has been found to vary by a kilometer or more. The troposphere has been characterized by falling temperature with increasing altitude, as the distance from heat-radiating earth increases. In the absence of air pollution the composition of the troposphere has been somewhat homogeneous. This homogeneity has been due largely to mixing by the constantly circulating air masses in the troposphere. The water content of the troposphere however, is extremely variable due to cloud formation, precipitation, and evaporation of water from terrestrial water bodies.

Table 2

Major Regions of the Atmosphere and their Characteristics

Region	Temperature Range, °C	Altitude range, km*	Significant chemical species
Troposphere	15 to —56	0 to 11	N_2, O_2, CO_2, H_2O
Stratosphere	—56 to —2	11 to 50	O_3
Mesophere	—2 to —92	50 to 85	O_2 + NO+
Thermosphere	—92 to —1200	85 to 500	O_2 +, O^+, NO^+

*These figures are values ranging from the lowest part of the region to the highest part. They may vary.

Troposphere. The zone nearest the earth, the troposphere, is named from the green *tropein* (to turn, to rotate, to change). One property that changes in the troposphere is the temperature. It drops to —55° C(—65°F) by the time an altitude of about 16-18 km (10-11 miles) is reached. The rate of temperature drop is about 6.4°C per 1000 metres (3.5°F per 1000 feet) a figure called the *normal environmental lapse rate*. The region through which this lapse rate occurs defines the troposphere. At the equator the troposphere is about 18 kilometers (11 miles) thick, but at the poles it is only about 8 kilometers (5 miles) thick. Essentially all the atmosphere's water vapour is in the troposphere which therefore encloses essentially all the storms and precipitation of the earth. The jet stream occurs at an altitude of 10.5-12 kilometers (35,000 - 40,000 feet; 6.5 - 7.5 miles).

The very cold layer at the top of the troposphere is known as the *tropopause*. Its low temperature and resulting condensation of water to ice particles prevents water from reaching altitudes where it would photo dissociate through the action of intense high energy ultraviolet light. If this were to happen, the hydrogen produced would escape the earth's atmosphere. Hence the tropopause acts as a barrier preventing the net loss of hydrogen from the earth with accompanying decomposition of water.

The weather patterns that we are familiar with are all contained within the troposphere, and mixing in the region is fast, of the order of weeks within a hemisphere, and 1 to 2 years between hemispheres. Vertical mixing occurs in the troposphere but movement into the stratosphere is inhibited by the temperature inversion.

Stratosphere. This zone starts at the top of the troposphere, at a thin atmospheric "shell" called the tropopause, where the temperature begins to stay fairly constant, increasing slightly, in going to an altitude of 50 kilometers (31 miles). The stratosphere includes much of the *ozone layer*. The warming trend in the stratosphere is caused by a cycle of chemical changes in the ozone layer which converts all the incoming high energy ultraviolet rays to heat. If these rays reached the earth's surface, life as we know it would be impossible. The *stratopause* marks the narrow zone at the top of the stratosphere where the temperature begins to fall again with increasing altitude as the mesosphere is entered.

The stratosphere is characterized by a rising temperature as the altitude increases. The temperature reaches a maximum at the upper limit of the stratosphere. The temperature increase of the stratosphere is due to the presence of ozone, which may reach a level of 10 ppm by volume in the mid-region of the stratosphere. Ozone absorbs energy in the form of ultraviolet light and causes an increase in temperature.

The stratosphere has been of interest to aeronautical scientists because it is traversed by airplanes; to communication scientists because of radio and television communications; and to air pollution scientists because global transport of pollution, particularly the debris above ground, atomic bomb tests and of volcanic eruptions, occurs in the stratosphere; and because absorption and scattering of solar energy also occurs in the stratosphere.

The principal chemical species in the stratosphere, N_2, O_2, O_3 and some H_2O, are chemically active because of their interaction with ultraviolet solar radiation. The interaction of a species with radiation, producing chemical changes, is *photochemistry*. Radiation is absorbed forming an activated species and, provided the conditions are suitable, a new species is formed.

Though most of the energetic radiation from the sun has been filtered out at higher altitudes, some important chemical reactions take place in the troposphere, particularly at the atmosphere/biosphere and atmosphere/lithosphere interfaces. Two of these reactions are photosynthesis, the overall process given by the two equations :

$$2H_2O \rightarrow O_2 + 4H + 4e^- \quad ..(1)$$

$$CO_2 + 4H+ + 4e^- \rightarrow [CH_2O] + H_2O \quad ..(2)$$

and nitrogen fixation:

$$N_2 \xrightarrow[\text{hydrogenase}]{\text{nitrogenase}} NH_2 \rightarrow NO_2^- \rightarrow NO_3^- \quad ..(3)$$

in the soil

Chemical reactions, for example, hydrolysis and

oxidation, also occur at the surface of the earth. The corrosion:

$$Fe—2e \rightarrow Fe^{2+} \quad ..(4)$$

$$O_2 + 2H_2O + 4e \rightarrow 4OH^- \quad ..(5)$$

in

i.e., $2Fe + O_2 + 2H_2O \rightarrow 2Fe^2 + 4OH \rightarrow Fe_2O_3$ x H_2O.

air

rust ..(5)

Electrical discharges in the atmosphere initiate reaction between nitrogen and oxygen as follows:

$$N_2 + O_2 \xrightarrow{hv} 2NO \xrightarrow{O_2} 2NO_2 \quad ..(7)$$

The farultraviolet radiation (wavelength <190 mm) is removed before it reaches the stratosphere by atomic nitrogen and oxygen. A considerable portion of longer wavelength ultraviolet, (i.e., below 340 nm) is removed in the stratosphere, for example, radiation of wavelength <242 nm promotes the reaction:

$$O_2 + hv \rightarrow O + O^* \quad ..(8)$$

The atomic oxygen produced reacts with O_2 to form ozone:

$$O^* + O_2 + M \rightarrow O_3 + M^* \quad ..(9)$$

and the ozone absorbs ultraviolet radiation of wavelength <340nm;

$$O_3 + hv \rightarrow O_3 + O \quad ..(10)$$

followed by the reaction:

$$O_3 + O \rightarrow 2O_2 \quad ..(11)$$

Hence a cycle of ozone formation and destruction (Chapman mechanism) exists i.e. $3O_2 \rightarrow 2O_3$ ..(12)

and $2O_3 \rightarrow 3O_3$ ..(13)

Other reactions can also occur, the principal among which is the reaction of ozone with nitric oxide.

$NO + O_3 \rightarrow NO_2 + O_2$ ..(14)

$NO_2 + O \rightarrow NO + O_2$ ..(15)

$NO_2 + hv \rightarrow NO + O\lambda < 400$ nm) ..(16)

The overall process may be put as before:

$2O_3 \rightarrow 3O_2$ [twice reaction (14) plus reactions (15 and (16)] ..(17)

and account for removal of approximately 70% ozone. Species HO_x are involved in similar reactions.

The absorption of ultraviolet radiation in the stratosphere prevents virtually all radiation below 290 nm, and most between 290 and 320 nm from reaching the earth. Present day biological life has been protected from this radiation, a situation that may not continue as gaseous pollutants are added to the atmosphere.

In reaction (9) an excited species, M^{*}, is produced. The energy may be transferred as heat, which would explain rise in temperature within the stratosphere. Infrared radiation also absorbed and re-emitted by ozone adds to the temperature rise.

Mesosphere. The mesosphere, on top of the stratosphere, extends roughly to 80 kilometres (50 miles), and its temperature drops to between—80 and 90°C at the mesopause, a thin zone where the temperature stabilizes and son starts to rise again as the *thermosphere* is ascended.

The principal chemical species in the mesophere are N_2, O_2, O_2^+ and NO^+, and in the thermosphere are N_2, O_2, O_2^+, O, O^+ and NO^+. The thermosphere is also called the ionosphere as ionization of molecular and atomic species takes place in the lower layers. The layers are responsible for the reflecion of radio waves.

Thermosphere. This zone is above 80 kilometres (50 miles) altitude. The temperature becomes very high, but the air has such a low density that it can hold very little heat. Any denser object in the thermosphere will be extremely hot in sunlight but very cold at night.

The temperature falls with increasing altitude in the mesosphere due to a decrease in the levels of the radiation—absorbing species, particularly ozone. In the higher altitudes of the mesosphere and above, molecules and atoms can completely escape the earth's atmosphere. (This region is called the exosphere). The temperature rises to a maximum of approximately 1200°C in the thermosphere due to the absorption of highly energetic radiation of wavelengths less than approximately 200 nm by species in this region.

Atmospheric pressure tends to decrease as an approximately exponential function of altitude. Ideally, in the absence of mixing and at a constant absolute temperature, T, the pressure at any given height, Ph, is given by the exponential equation

$$P_h = P_0 e^{-Mgh/RT} \quad ..(1)$$

Where P_0 refers to the pressure at zero altitude (sea level), M the average gram molecular weight of air (28.97 grams/mole), *g* the acceleration of gravity (981 cm sec^{-2} at sea level), *h* the altitude in cm, and R the gas constant (8.314 x 10^7 erg deg^{-1} mole^{-1}). For consistency the units in Equation (1) are given in the c.g.s. system, although height is normally measured in metres or kilometres and it becomes necessary in using Equation (1) to convert height to cm by multiplying by the appropriate factors.

The pressure drops by the factor e—1 for each increase in altitude equal to the *scale height*, defined as RT/Mg. At an average sea level temperature of 288°K the scale height is 8 x 10^5 cm or 8 km. At an altitude of 8 km the pressure is only about 39% of sea level pressure.

In order to get a better of the variation of pressure with altitude. Equation (1) may be converted to the logarithmic (base 10) form and *h* expressed in kilometres

$$\log P_h = \log P - \frac{Mgh \times 10^5}{2.3.3RT} \quad ..(2)$$

If the pressure is taken as 1.00 atm at sea level, the equation becomes simply

$$\log P_h = -\frac{Mgh \times 10^5}{2.3.3RT} \quad ..(3)$$

In terms of bulk, by far the greatest portion of the atmosphere is confined to the region very close to the surface of the earth. Less than 1% of atmospheric gases are more than 40 km from the earth's surface. Such an altitude is of course miniscule compared to the radius of the earth.

The zones just discussed may be defined by changes of temperature with altitude. Changes in chemical composition or activity also occurs with altitude. Up to about 80 kilometers (50 miles) the proportions of components given remain nearly constant, and the long zone up to that altitude is sometimes called the *homosphere*. Within it is a region called the *chemosphere* extending from the upper troposphere where a huge number of chemical changes occur powered by solar radiation. Ozone production is one such activity. Above the homosphere lies the *heterosphere*. For its first 40 kilometers (25 miles), the ionosphere, the very thin concentration of atoms and molecules, exists largely as electrically charged particles called ions. This globe-enveloping band of ions reflects outgoing radio waves back to earth, making radio transmission possible well beyond the horizon. Television waves are not reflected, and a TV transmiter's range does not extend beyond the horizon. Meteorities entering the earth's atmosphere are generally out in the lonosphere.

Particles in the Atmosphere

Particles constitute common and significant components of the atmosphere, especially the trosposphere. Colloidal-sized particles in the atmosphere are called aerosols. Most aerosols from natural sources are having a diameter of less than 0,2 microns. These particles are termed as Aitken particles and are noted for their ability to induce formation of water droplets in supersaturated atmospheres. Aitken particles originate in nature from sea sprays, smokes, dusts, and the evaporation of organic materials from vegetation. Other typical particles of natural origin in the atmosphere include bacteria, fog, pollen grains, and volcanic ash.

Among the atmospheric phenomena which involve aerosol particles there have been electrification phenomena in the atmosphere, cloud formation, and fog formation. Particles help determine the heat balance of the earth's atmosphere by reflecting light. Probably the most important function of particles in the atmosphere has been their action as nuclei for the formation of ice crystals and water droplets. Current efforts at rain making are centred around the addition of condensing particles to atmospheres supersaturated with water vapour. Dry ice was used in early attempts; now silver iodide, which forms huge numbers of very small particles, is employed.

Particles are involved in many chemical reactions in the atmosphere. Neutralization reactions, which take place most readily in solution, may occur in water droplets suspended in the atmosphere. Small particles of metal oxides and carbon are having a catalytic effect on oxidation reactions. Particles may participate in oxidation reactions which are induced by light. It has been shown that zinc oxide surfaces, photosensitized by light, may cause the oxidation of water to hydrogen peroxide.

Particles may vary in number from several hundred per cubic centimeter, in very pure air, to more than 10^5 per cubic centimeter in highly polluted air. Their size lies in the range of 0.1 —10m. The particulate mass level has been found to vary from 10 gm/m^3 in clean air to 60-2000 gm/m^3 in the polluted air in urban areas.

In USA nearly 13.4 MT (metric tonnes) of primary particles have been given out of which about 5 MT have been smaller than 3m.

Metal oxides constitute a major group of inorganic particles in the atmosphere. They are originated from the combustion of fuels. Thus, particulate iron oxide is originated in the combustion of pyrite—containing coal :

$$3Fe_2 + 8O_2 \rightarrow Fe_3O_4 + 6SO_2 \quad ..(3)$$

A part of the $CaCo_3$ in the ash fraction of coal charged to CaO and gets discharged through the stack :

$$CaCO_3 + heat \rightarrow CaO + CO_2 \quad ...(4)$$

Automobiles have been the main sources of lead particles in the atmosphere. Combustion of leaded gasoline releases lead halides through the exhaust system

$$Pb(C_2H_5)_4 + O_2 + \text{(halogenated scavengers)} \rightarrow PbCl_2$$

+

$$PbBr_2 + PbBrCl + CO_2 + H_2O \qquad ..(5)$$

Aerosol mists are formed from sulphuric acid which are obtained by oxidation of sulphur dioxide which collects water vapour thereby forming small liquid droplets :

$$2SO_2 + O_3 + 2H_2SO_4 \qquad ..(6)$$

Chemical Processes for Formulation of Organic Particulate Matter

Organic particulate matter are formed from a wide variety of sources such as emissions from vegetation and automobiles, combustion of fuel, etc. Typical organic particulate matter, which is obtained in the form of a benzene-extractable fraction from 200 air samples in USA reveals an average formula, $C_{32.4'}$ H_{48} $O_{3.8}$ $S_{0.083}$ $halogen_{0.065}$ $alkoxy_{0.12}$. Such particles have been in the 1m size range and have been regarded as potential health hazards.

Polycyclic (polynuclear) aromatic hydrocarbons (PAH) present in organic particulate matter are having carctnogenic effect. Among these, benzo (α-) pyrene is a well characterized substance. PAH compounds commonly are found in urban atmospheres at about the $20\mu g/m_3$ level. They are obtained from the pyrolysis of higher paraffins present in fuels and plant material.

High molecular weight paraffins are pyrolysed for getting $C_{10}H_{32}$ which is further pyrolysed. Most of the PAH compounds get absorbed on soot particles.

Soot is obtained as a residue on combustion of fuel in power plants and automobiles. It constitutes nearly 50% of the particulate load in urban areas. Chemically, it refers to a highly condensed product of polynuclear aromatic hydrocarbons; having several thousand interconnected crystallites, i.e., graphitic platelets each possessing nearly about 100 condensed aromatic rings. The hydrogen content of soot has been 1-3% and oxygen content about 5-10% due to partial surface oxidation. Due to its large surface area, soot is known to act as a carrier for toxic organics, *e.g.,* benzo (a) pyrene, and toxic trace metals, *e.g.*, beryllium, cadmium, chromium, manganese, nickel, vanadium, etc.

In general, atmospheric particles enter the respiratory tract of human bodies. However, small parties (0.1-1μ size) reach the lungs and are retained there, bringing about health hazards. The removal of particulate matter from gas streams is regarded to be an important strategy for air pollution control.

Chemical and Photochemical Reactions in the Atmosphere

One of the primary obstacles encountered in studying atmospheric chemistry is that the chemist generally must deal with incredibly low concentrations. The detection and analysis of reaction products is quite difficult. Simulating high altitude conditions in the laboratory can be extremely hard due to interferences, such as those from species given off from container walls under conditions of very low pressure. Many chemical reactions which need a third body to absorb excess energy occur very slowly in the upper atmosphere where there is a sparse concentration of third bodies, but take place readily in a container whose wall effectively absorbs energy. Container walls may serve as catalysts for some important reactions or they may absorb important species and react chemically with the more reactive ones.

The absorption of light by chemical species can cause reactions which do not otherwise take place under the conditions (particularly the temperature) of the medium in the absence of light. thus photochemical reactions, even in the absence of a chemical catalyst, take place at temperatures much lower than those which otherwise would be required. Such photochemical reactions, induced by intense solar radiation, play a very important role in ascertaining the nature and ultimate fate of a chemical species in the atmosphere.

Electronically excited molecules have been one of the three relatively reactive and unstable species which are encountered in the atmosphere and which are strongly involved with atmospheric chemical processes. The other two species are atolms or molecular fragments with unshared electrons, called free radicals and ionized atoms or molecular fragments.

Electronically excited molecules are produced when stable molecules absorb energetic electromagnetic radiation in the ultraviolet or visible region of the spectrum. A molecule may possess several possible excited states, but generally ultraviolet and visible radiation are energetic enough to excite molecules only to several of the lowest energy levels. The nature of the excited state may be understood considering the disposition of electrons in a molecule. Most molecules have an even number of electrons. The electrons occupy orbitals, with a maximum of two electrons having opposite spin occupying the same orbital. The absorption of light may promote one of these electrons to a vacant orbital of higher energy.

These excited states are energized compared to the ground state and are chemically reactive species.

Two conditions must be met before a photochemical reaction will occur. The first of these lawqs of photochemistry is that light must be absorbed by the reacting species. If the absorbed light is in the visible region of the sun's spectrum, the absorbing species is coloured. Coloured NO_2 is the prime example of such a species in the atmosphere. The second law of photochemical process involves activation of the molecule by the absorption of one quantum of light. The energy of one quantum is equal to the product *hv*, where h is Planck's constant, 6.62 x 10^{-27} erg sec, and *v* is the frequency of the absorbed light, sec^{-1}.

An intermolecular energy transfer may take place in which an excited species transfers energy to another species. A subsequent reaction by the second molecule is termed as a photosensitized reaction. It is believed that part of airglow arises from luminescence by excited sodium atoms, Na^*, formed in the atmosphere by an intermolecular energy transfer involving the following reaction:

$$O_2^* + Na \rightarrow O_2 + Na^*$$

In some cases only energy transfer within the molecule occurs,

$$XY^* = XY+$$

a process called intramolecular transfer, where the dagger, +, denotes an excited state different from the initial one. Finally, the excited species may undergo spontaneous isomerization, as in the conversion of O-nitrosobenzaldehyde to O-nitrosobenxoic acid, a reaction used in chemical actinometers to measure exposure to electromagnetic radiation.

The absorption of very energetic radiation may result in the detachment of an electron,

$$N_2 + hv = N_2^+ + e$$

a process called photoionization. Photoionzation often is considered a subcategory of photodissociation in which one of the dissociation products is an electron.

Light absorbed in the infrared region is not

sufficiently energetic to break chemical bonds. The receptor molecules gain vibrational and rotational energy. The energy absorbed as infrared radiation ultimately is dissipated as heat and raises the temperature of the atmosphere. The absorption of infrared radiation is very important in the earth's acquiring heat from the sun and in the retention of energy radiated from the earth's surface.

Reaction of Atmospheric Nitrogen

The atmosphere contains 78% nitrogen by volume and constitutes an inexhaustible reservoir of that essential element.

Unlike oxygen, which is almost completely dissociated to the monatomic form in higher regions of the thermosphere, molecular nitrogen is not readily dissociated by ultraviolet radiation. However, at altitudes exceeding approximately 100 km atomic nitrogen is produced by photochemical reacts;

$$N_2 + hv \rightarrow N + N \quad ..(1)$$

Othe reactions which may produce monatomic nitrogen have been the following:

$$N_2^+ + O \rightarrow NO^+ + N \quad ..(2)$$

$$NO^+ + e \rightarrow N + O \quad ..(3)$$

$$O^+ + N_2 \rightarrow NO^+ + N \quad ..(4)$$

In oe region of the ionosphere, the so-called E region, NO^+ has been one of the predominations. A plausible sequence of reactions by which NO^+ is formed as the following:

$$N_2 + hv = N_2^+ + e \quad ..(5)$$

$$N_2^+ + O = NO^+ + N \quad ..(6)$$

In the lowest region of the ionosphere, the D region

which extends from approximately 50 km in altitude to approximately 85km, NO^+ is formed by ionizing radiation:

$$NO + hv = NO^+ + e \quad ..(7)$$

In the lower part of the d region the ionic species N_2^+ is formed through the action of galactic cosmic rays:

$$N_2 + hv = NO^+ + e \quad ..(8)$$

Pollutant oxides of nitrogen, particularly NO_2. are key species involved in air pollution and the formation of photochemical smong. For example, NO_2 is readily dissociated photochemically to NO and reactive atomic oxygen:

$$NO_2 + hv = NO + O \quad ..(9)$$

This reaction is the most important primary photochemical process involved in smong formation.

Reactions of Atmospheric Oxygen

The primary features of the interaction of oxygen among the atmosphere, lithosphere, hydrosphere. The oxygen cycle is critically important in atmospheric chemistry, geochemical transformations, and life processes.

Oxygen in the troposphere plays a vital role in processes on the earth's surface. Atmospheric oxygen takes part in energy producing reactions, like the burning of fossil fuels.

$$CH_4 \text{ (in natural gas)} + 2O_2 = CO_2 + 2H_2O \quad ..(1)$$

Atmospheric oxygen is used by aerobic organisms in the degradation of organic material.

$$CH_2O + O_2 \xrightarrow{\text{organisms}} CO_2 + H_2O \quad ..(2)$$

Some oxidative weathering processes, like

$$4FeO + O_2 \rightarrow 2Fe_2O_3 \quad ..(3)$$

consume oxygen. Oxygen gets returned to the atmosphere through the photosynthetic reactions of plants:

$$CO_2 + H_2O \xrightarrow{\text{plants}} CH_2O + O_2 \quad ..(4)$$

Due to the extremely rarefiex atmosphere and the effects of ionizing radiation,. oxygen in the upper atmosphere exists in some forms which are quite different from those stable at lower levels. In addition to molecular oxygen, O_2, the upper atmosphere is having oxygen atoms, O; excited oxygen molecules, O_2^*; and ozone, O_3.

Atomic oxygen, O, is stable primarily in the thermosphere where the atmosphere has been so rarefied that the three-body collisions necessary for the chemical reaction atomic oxygen seldom take place. Atomic oxygen is produced by a photochemical reaction:

$$O_2 + h\upsilon \rightarrow O + O \quad ..(5)$$

Ultraviolet radiation in the wavelength regions 135 nm to 176 nm, and 240 nm to 260 nm has been most effective in causing reaction (5) to occur. Due to photochemical dissociation, molecular oxygen, O_2, is virtually nonexistent at very high altitudes. At altitudes exceeding about 80 km the avrage molecular weight of air is lower than the 28.97 g/ mole observed at sea level due to the high concentration of atomic oxygen resulting from photochemical dissociation of O_2. The resulting division of the atmosphere between a lower section with a uniform molecular weight and a higher region with a non-uniform molecular weight forms the basis for classifying two atmospheric regions as the homosphere and heterosphere, respectively. less than 10% of the oxygen in the atmosphere, at altitudes exceeding approximately 400 km is present as O_2.

Atomic oxygen is partially responsible for the phenomenon known as airglow. Excited atomic oxygen, O^*, given out visible light at wavelength of 636 nm, 630 nm and 558 nm. This species is formed by reactions such as the following:

$$O + O + O \rightarrow O_2 + O^* \quad ..(6)$$

Airglow is a very faint electromagnetic radiation which is continuously emitted by the earth's atmosphere. Although the visible component of airglow is extremely weak, the phenomenon is quite intense in the infrared of the spectrum.

Oxygen ions, O^+ may be formed by ultraviolet radiation acting upon oxygen atoms:

$$O + h\nu \rightarrow O^+ + e \quad ..(7)$$

O^+ is the predominant positive ion in some regions of the ionosphere. it may undergo the following reactions to form other inportant positive ions:

$$O^+ + O_2 \rightarrow O_2^+ + O \quad ..(8)$$

$$O^+ + N_2 \rightarrow NO^+ + N \quad ..(9)$$

In intermediate regions of the ionosphere, the species O_2^+ is formed by absorption of ultraviolet radiation at wavelengths of 17 nm to 103 nm. The reaction

$$O_2 + h\nu = O_2^+ + e \quad ..(10)$$

may also be energized by soft X-rays. Reaction (8) and the

$$N_2^+ + O_2 = N_2 + O_2^+ \quad ..(11)$$

reaction also give rise to O_2^+ in this region.

Ozone, O_3, is an extremely significant oxygen containing species occurring in the stratosphere. Ozone absorbs harmful ultraviolet radiation and serves as a radiation shield protecting living beings on the earth from the effects of excessive amounts of such radiation. Ozone is formed by a photochemical reaction.

$$O_2 + hv \rightarrow O + o \qquad ..(12)$$

(where the wavelength of the exciting radiation must be less than 242.4 nm) followed by a three-body reaction,

$$O + O_2 + M \rightarrow O_2 + M \qquad ..(13)$$

in which M is another species, like a molecule of N_2 or O_2, which absorbs the excess energy given off by the reaction and enables the ozone molecule to stay together. The region of maximum ozone concentration occurs within the range of 25 to 30 km high in the stratosphere. In this region the ozone concentration may reach 10 ppm.

Ozone absorbs ultraviolet light very strongly in the region of 220 nm to 330 nm. If this light were not removed by ozone, severe damage would cause to exposed forms to life on the earth. Absorption of electromagnetic radiation by ozone is also responsible for the temperature maximum encountered at an altitude of approximately 50 km, the boundary between the stratosphere and the mesophere. The reason that the temperature maximum takes place at a higher altitude than that of maximum ozone concentration is that ozone is such an effective absorber of ultraviolet light that only a relatively small fraction of the ultraviolet light reaches the lower altitude. One of the primary environmental problems resulting from the high altitude flights of supersonic planes has been that the possibility that exhaust products from these aircrafts cause reactions removing ozone from the atmosphere. Freon from aerosol cans might be having the same effect.

The absorption of ultraviolet light may bring about the dissociation of ozone.

$$O_3 + h\nu \rightarrow O_2 + O \qquad ..(14)$$

or ozone may decompose upon reaction with atomic oxygen :

$$O_3 + O \rightarrow O_2 + O_2 \qquad ..(15)$$

Atomic oxygen may also be eliminated by the following reaction :

$$O + O + M \rightarrow O_2 + M \qquad ..(16)$$

Atmospheric Carbon Dioxide

Carbon dioxide is composing only 0.0314 per cent by volume of the atmosphere. It is the non pollutant species of most concern in the atmosphere. Carbon dioxide, along with water vapour, is mainly responsible for the absorption of infrared energy from the sun, and radiation of this energy back to the earth's surface. Now, scientists fear that changes in the atmospheric carbon dioxide level will bring about marked changes in the earth's climate resulting in global warming.

In terms of strictly chemical reactions in the atmosphere, carbon dioxide is a relatively insignificant species. It is present in relatively low concentrations and is not thought to participate significantly in photochemical reactions. The infrared radiation which it absorbs is not sufficiently energetic to cause chemical reactions.

Water in the Atmosphere

The water vapour content of the atmosphere has been to

vary over a wide range, especially in the lower atmosphere. The normal range has been 1 to 3 per cent of on a volume basis, although air may be having as little as 0.1 per cent of water and as much as 5 per cent. Water circulates through the atmosphere in the hydrologic cycle.

Water vapour absorbs infrared radiation, even more than carbon dioxide, thereby determining the heat balance of the earth. Clouds formed from water vapour are able to reflect light from the sun and have a temperature lowering effect.

Gaseous water in the atmosphere is not known to participate appreciably in most atmospheric chemical reactions. However, photochemical reaction between water and ultraviolet light—

$$H_2O + hv = H + OH \quad ..(1)$$

is regarded to be a major contributor of atomic hydrogen in the atmosphere. Excited hydroxide radicals, OH*, are formed by the reaction of atomic hydrogen with ozone.

$$H + O_3 \longrightarrow OH^- + O_2 \quad ..(2)$$

The radicals infrared radiation which ranges into the extremely long wavelength regions of the visible spectrum :

$$OH^* = OH + hv \quad ..(3)$$

This emission has been partially responsible for the phenomenon of airglow.

Condensed water vapour in the form of very small droplets has been of considerable significance in atmospheric chemistry. Pollutant sulfur dioxide gets oxidized to sulfur trioxide in the atmosphere.

$$2SO_2 + O_2 = 2SO_3 \qquad ..(4)$$

and the sulfur trioxide gets incorporated into droplets of water—

$$SO_3 + H_2O = H_2SO_4 \qquad ..(5)$$

to form sulfuric acid. In some industrial areas the rainfall may have a remarkably low pH as a consequence. The harmful effects of some air pollutants—for instance, the corrosion of metal; by acid forming gas—required the presence of water which may come from the atmosphere. The presence of water vapour has an important influence upon pollution-induced fog formation under some circumstances. Water vapour interacting with particular matter in the atmosphere may reduce visibility to undesirable levels through the formation of aerosol particles.

Fog

Fog forms in the atmosphere close to the earth's surface when the relative moisture content reaches 100%, that is when the vapour pressure equals or exceeds saturation point. As the physical processes of cloud formation and fog formation are identical, fog can also be described as 'clouds close to the ground.'

Ground fog is formed in the evening and at night when the radiation balance of the earth is negative, that is, when outward radiation from the earth exceeds inward radiation from the sun. By day the atmosphere near the earth's surface is charged with water vapour, which because of the sharp drop in temperature at night condenses. Fog formation is intensified when, 'owing to a flow of cold air in the valleys, cold air accumulates or when the water content of the ground (in marshland, wet meadows, etc.) is very high. As ground fog occurs in "radiation nights" and is always associated with temperature inversion in the atmosphere near ground level, it is often referred to as "radiation fog."

Mainly in the autumn an inflow of cold air spreads across warm water or warm moist earth and gives rise to *dank fog*. As in such areas the temperature differences between a cold air and warm water may amount to 10°C

(50°F) or more the surface water evaporates quickly and the water vapour condenses again immediately in the cold air. The fog so formed breaks rapidly, owing to the unstable temperature layers (warm below cold) and takes on the appearance of smoke. A similar pattern is observed in the summer over woods and roads when the sun comes out again. The water evaporates owing to the powerful energy input on the woodland and road surfaces, and condenses again immediately in the rising air.

We speak of *mixing fog* (warm front fog) when warm air is suspended above a layer of cold air close to the ground so that, owing to the interchange and turbulence of the water vapour, the warm air can generate into the cold air below it. As a result of the drop in the temperature of the mixture caused by intermingling in relation to the saturated warm air the saturated moisture pressure will be reduced and the water vapour condensed.

Owing to thermal reaction in the atmosphere under low-lying inversions, *high level* fog is formed. Often it extends downwards and so becomes ground fog. Although other factors may contribute to its formation, high level fog is primarily a reduction fog.

Advection fog forms; when warm air of sub-tropical origin extends over a colder layer of air in the warm air zone of a depression. In this case, the warm air is cooled down by a turbulent exchange from ground level upward, so that upon reaching the point of condensation, fog is formed. Such fog formation occurs only in the winter months, when the temperature differences between the sub-tropical and the temperature zone are substantial. In the summer the differences are too small to promote any noteworthy cooling off. Fog formation of this order also occurs when warm moist air passes over cold water surfaces. This phenomenon is common to all cold marine current and occurs mainly in the summer.

Temperature of the Air Layer Near the Ground

A factor of far-reaching importance of the life processes in the biosphere is the temperature. The temperature, as measured by changes in the volume of mercury in a thermometer, is a reflection of the average kinetic energy in a molecule of air, which is derived essentially from the impact of solar radiation on the earth's surface. The temperature of the layer of air nearest the ground is thus determined by the extent of the solar energy radiated. This in turn is influenced by the composition of the atmosphere, the geographical latitude, the nature of the soil or rock surface, the features of the terrain, etc. The inflow of cold or hot air masses from extensive currents can change the temperature of the layer of air near the ground very substantially.

By day part of the sun's rays penetrate the atmosphere to reach the earth's surface, where some are coverted into thermal radiation (infrared rays) and returned to the atmosphere. This thermal radiation is absorbed by the atmosphere, which thus becomes richer in energy, i.e., warmer. As most of the energy is imparted to the molecules nearest to the source of energy, that is, the earth, the air immediately above the ground is warmed most. In the temperature zone on a fine summer day with no wind, temperatures of over 21°C (70°F) are recorded in the atmosphere close to the ground. The warm air close to the earth's surface is lighter in specific gravity than the cold air above it, and consequently it tends to rise. During the upward flow, the energy rich molecules impart energy to the surrounding atmosphere, until an energy balance is achieved. Owing to the radiation of solar energy by day, a temperature gradient forms in the layer or air close to the ground. The air temperature falls as the height above the earth's surface increases. This process always occurs during radiation, about, is naturally more marked when air movements are slight. Strong air movements change the temperature conditions in the layer of air near the ground considerably.

At night the thermal radiation from the soil or rock surface of the earth is no longer replenished by solar radiation. As the energy source is lacking, radiation during the night weakens progressively. Consequently the cold air descending towards the ground is not warmed up again and remains as a layer of cold air on the earth's surface. According to the nature of the terrain, the cold air will flow under the pull of gravity into depressions such as valleys, ravines and sink holes to form cold air pockets. Temperature differences of 10°C (50°F) and more between cold air pockets and the warmer zones at higher levels are by no means rare.

Clouds reduce the radiation of the thermal energy considerably, by deflecting it form their underside. Sudden temperature increases of 50°C (44°F) and more observed in the vicinity of a cloud bank. Strong winds also obstruct the formation of a temperature gradient at night. The temperature gradient in the layer of air near the ground is consequently subject to constant reversal between day and night. The conditions that lead to the formation of these gradients can be affected by wind, by the character and structure of the terrain, and by cloud formation and other atmospheric influences.

Evolution of the Atmosphere

It is now widely accepted that the earth's atmosphere originally was very different from its present state, and that the changes were brought about by biological activity.

Nearly 3.5 billion years ago when the first primitive life molecule got formed, the atmosphere was chemically reducing, having primarily methane, ammonia, water vapour, and hydrogen. The atmosphere was bombarded by intense, bond-breaking, ultraviolet light. This light, along with lightening and radiation from radionuclides, provided the energy to bring about chemical reactions which caused the production of relatively complicated molecules, including even amino acids and sugars.

From the rich chemical mixture in the sea, life molecules got evolved. Initially these very primitive life forms derived their energy from fermentation of organic matter formed by chemical and photochemical processes, but eventually they gained the capability to produce organic matter, "CH_2O", by photosynthesis.

$$CO_2 + H_2O + h\nu \rightarrow CH_2O + O_2 \quad ..(1)$$

and the stage was set for the massive biochemical transformation which caused the production of all the atmosphere's oxygen.

The life forms might have been blue-green algae whose limy deposits have been occurred in the oldest of Pre-Cambrian rocks. They were assumed to be first authentic members of the plant world (1400 million years ago). Probably, there was some oxygen which was essential to support the primitive life form. This presumably got formed in the upper atmosphere by the breakdown of water vapour by electrical discharge and solar radiation. The first plant and similar species might be the earth's first mass-producers of free O_2. When the supply of O_2 got increased more complex plants got evolved and accelerated the process which got stepped up for millions of years, accompanied by an enormous increase in O_2—consuming animals to help maintain the O_2—balance. The present composition of the atmosphere has been virtually the same as it was in the Cambrian age (500 million years ago). But it has been continually changing. The mobiles gases of which it has been composed are always being absorbed by the soil, the rocks and living things or lost into outer space. The supply of atmospheric CO_2— is to be replenished every 4 to 8 years, O_2 every 3000 years and N_2' every 100 million years. The cycle of use and renewal of the atmospheric gases depicts an intriguing balance involving soil,air, animals and plants. It is to be noted that all living things have been mainly made of the elements of air; O_2' H_2O, C and N.

The oxygen initially produced by photosynthesis was probably quite toxic to primitive life forms. However, much of this oxygen got converted to iron oxides,

$$4Fe^{2+} + O^2 + 4H_2O = 2Fe_2O_3 + 8H^+ \quad ..(2)$$

by reaction with soluble iron (II). This caused the formation of enormous deposits of iron oxides, which provide major evidence for the liberation of free oxygen in the primitive earth atmosphere.

Eventually enzyme systems developed which enabled organisms to mediate the reaction of waste product oxygen with oxidizable organic matter in the sea. Later this mode of waste product disposal was used by organisms for producing energy by respiration, which is now the primary mechanism by which most organisms obtain energy.

Eventually oxygen got accumulated in the atmosphere. This of course provided an abundant source of oxygen for respiration. It was having an additional benefit in that it enabled the formation of an ozone shield. This ozone shield absorbs bond-rupturing ultraviolet light. With the ozone shield protecting tissue for destruction by high energy ultraviolet radiation, the earth became a much more hospitable environment for life, and life forms were enabled to move from the sea to land.

7 Natural Environment and Management

The perception that man lives in close relationship with all living things, plants and animals, is quite old. Ancient and medieval philosophers have often emphasized the inter-relatedness of the entire living world. Recent developments in science have not only confirmed this view but have gone beyond to assert the inter-dependence of all life, microbial, plant and animal, between themselves and with the physical environment of land, air, water and solar energy, this inter-dependence is embedded in the evolution of life during the past three billion years or more with the origin of various forms of life, the extinction of others, and the formation of our atmosphere, water bodies and land masses, these ideas have been sharpened in the last nearly fifty years by the concept of ecosystem (1935) which defines an area in which all living creatures and the soil-water-air-environment of this physical region of the earth have developed various inter-relations to produce a quasi-stable or almost stable system of mutual interdependence. The science of ecology is the study of such system, their inter-relationship and normal slow secular changes of such systems.

There is an even broader idea proposed by some scientists like Lovelock who would like to think of the entire biosphere as a single ecological system which he called Gaia. Conceptually, this view is reasonable and exciting and recalls the jain concept of 'Prakriti'. However, the current scientific methods of study are unable to deal with so large and complex a total system. Ecologists, therefore, usually define much smaller systems which are amenable to study, investigation and analysis. Some of these systems may be relatively self-contained and small enough to enable a scientific study to be pursued credibly. A forest, a lake, a valley or a river basin are examples of almost self-contained ecosystems. Normally the migration into almost self-contained systems, where it takes place is restricted both in the numbers of intruding species and in the absolute number of such species. There are other systems which may be more open, the analysis of such open systems is more difficult, Examples of such open systems are oceans and large open plains such as the Indo-Gangetic plains. However, certain biogeochemical processes are almost world-wide and their cycles represent an essential part of natural processes occurring within the more restricted area defined by an ecological system.

The Carbon and Oxygen Cycles

The bulk of living matter is made up of four chemical elements—carbon, hydrogen, oxygen, and nitrogen. For life to continue we have seen that these critical elements must move in interconnected chemical cycles throughout the biosphere. Although there are a number of important biogeochemical cycles, we shall look at three representative ones: the carbon, oxygen, and nitrogen cycles.

Carbon undergoes a complete cycle. It moves from CO_2 in the air and water to plants through photosynthesis, and from there through the food chain to plant-eating animals (herbivores) and flesh-eating animals (carnivores). At each step, part of the carbon stored in complex food molecules is broken down to provide energy for the organism and returned to the air and water as CO_2 by the process of cellular respiration. The remaining carbon is also returned to the air and water when the plant or animals eventually dies and is broken down into CO_2 by the bacteria and fungi that decompose organic matter. While the carbon is used over again, we see in Figure that the solar energy used to drive the cycle is converted first to chemical energy in the chemical bonds in the larger carbon containing molecules, and then primarily to heat in the atmosphere when the chemical energy is released as the molecules are broken down. This heat energy then flows back into space. The oxygen cycle is interdependent with the carbon cycle. When CO_2 and H_2O are used in photosynthesis by plants, O_2 is produced, and it is then converted back to H_2O and CO_2 by respiration in plant and animal cells.

Photosynthesis:

Carbon dioxide+water	solar → energy	large organic molecules	+oxygen
CO_2+H_2O	solar → energy	large organic molecules	$+O_2$

Cellular Respiration:

large organic molecules	+ oxygen	carbon → dioxide	+water+energy
large organic molecules	$+O_2$ →	CO_2	$+H_2O+$ energy

By stripping away some of the details, we can see more celery how carbon and oxygen are cycled while energy from the sun flows through the biosphere and is degraded to less useful forms in the process. The energy is converted from solar energy-chemical energy-heat. Plants convert the solar energy to chemical energy through photosynthesis. Some of this chemical energy is used by the plant or animal in respiration and other processes necessary for life. Because of the Second Law of Thermodynamics, the efficiency of energy conversion at each step is very low and most of it ends up as essentially useless heat energy in the atmosphere.

All of the carbon is not immediately recycled by the photosynthesis-respiration cycle in living organisms. In the course of millions of years part of the carbon from decayed plants and animal has been temporarily removed from the normal carbon cycling and incorporated by geological processes into the earth's crust as fossil or carbon fuels (coal, natural gas, and oil). These fossil fuels and rock represent a temporary storage of carbon and solar energy in concentrated form. As man burns, or combusts, these fossil fuels this carbon is returned to the carbon cycles as CO_2 and the solar energy temporarily stored as chemical energy is released back into the atmosphere as heat.

Since these fossil fuels took millions and millions of years to form, we must consider them as nonrenewable resources; once used this concentrate form of useful energy will no longer be available. The carbon tied up in carbonate rock formations is also eventually returned to the carbon cycle as CO_2 as these rocks undergo very slow dissolution by watering.

Now man has the potential to disrupt the carbon cycle on a local or global level. Since the industrial revolution we have been using fossil fuels at an increasing rate and thus releasing carbon back into the atmosphere as carbon dioxide, CO_2, and carbon monoxide, CO. Of course plants and animals and other natural processes release CO_2 into the atmosphere as part of the carbon cycle; but there is concern that man might eventually overload the natural carbon cycle. Our atmosphere has evolved over millions of years by the action of plants and animals. As a result of this it contains a balance of gases such as N_2, O_2, CO_2, and H_2O that supports life as we know it. Upsetting our present steady state, by releasing too much CO_2 into the atmosphere over a relatively short period of time (several hundred years), could have serious consequences.

During the last 100 years the carbon dioxide, CO_2 content of our atmosphere has risen almost 10 percent because of the burning of fossil fuels, and the rate is increasing. As a result, the CO_2 content of the atmosphere has increased at 0.2 percent each year since 1958. At this rate, by A.D 2000, the CO_2 content would be 25 percent higher than that.

It is believed by some that this rise in CO_2 content could eventually produce what is known as the "greenhouse effect". The CO_2 admits radiant energy radiation (heat) from escaping back into space. As CO_2 builds up there could be a gradual warming of the earth's atmosphere, perhaps as much as 0.5°C by the year 2000. A rise of only 2 to 3°C could cause the Antarctic ice sheet to melt or slip into the ocean, thus raising the sea level by any where from 100 to 400 feet, over a long period of year. Such a rise in the sea level would put most of the world's major cities under water-hardly a desirable "solution" to the overpopulation problem.

Such projections, however, are quite uncertain because we know so very little about our atmosphere. There is considerable debate in the scientific community as to whether this or other processes are occurring as well as debate on their long-term effects. Indeed, some project that the atmosphere may be cooling rather than heating because of increased amounts of dust from volcanoes and wind storms coupled with the dust, soot, and other particulate matter man is pouring into the atmosphere. It is hypothesized that a rise in particulate matter in the atmosphere would reflect or absorb some of the incoming solar radiation and thus prevent some of it from getting through the atmosphere, thus producing a cooling effect and possibly a new ice age. Such effects are probably not immediate dangers, but on a long-range basis they could be much serious. The important point is that we don't really know enough to predict what might happen—but we do know that for the first time man now has the potential to bring about changes in atmospheric balance on a local (for example, around a city) and a global scale. A recent authoritative study of these and other major environmental problems stated:

> "Although we conclude that the probability of direct climate change in this century resulting from CO_2 is small, we stress that the long-term potential consequences of CO_2 effects on the climate or of societal reaction to such threats are so serious that much more must be learned about future trends of climate change."

We can also potentially upset the oxygen and carbon cycles by decreasing photosynthetic activity through continuing elimination of plants and trees. In the United States we are bulldozing, chopping, and paving over our land with asphalt and concrete at an alarming rate—the equivalent to paving over the state of Rhode Island every six months.

We usually think of the familiar green plants and trees carrying out all photosynthesis and providing our oxygen, but at least one-third of all photosynthesis is carried out by the "grass of the sea," tiny microscopic plants called photosynthesis found floating in our oceans, rivers, and lakes. Man may already be upsetting the ecological balance in man of our water ecosystems. Pollution of the world's oceans is one of our most serious problems and it is projected to increase seven fold before A.D. 2000. We are presently dumping over 5,00,000 different man-made chemicals into or oceans, rivers, and lakes; and only a handful have been tested to determine their effect of photosynthesis.

Does this mean that we are about to run out of oxygen as many horror stories in the popular press have indicated? Fortunately, recent calculations indicate that we are in no real danger of running out of oxygen from these intrusions into the carbon and oxygen cycles. The danger lies more in upsetting the much needed food-producing capacity of our land and oceans by removing vegetation and disrupting natural checks and balances between various living organisms in land and water systems. More and more prime farm land is taken over by urbanization. As land is cleared much of the valuable nutrients are washed away into water systems—thus providing a nutrient leak on the land and is cleared much of the valuable nutrients are washed away into water systems—thus providing a nutrient leak on the land and frequently a nutrient over load in our rivers and lakes. It is these intrusions into the vital chemical cycles that most concern us.

Nitrogen is another element required by all living organisms. Indeed it occupies a special place in terms of human life because it is a sensitive indicator of the quality of health in humans. Nitrogen deprivation is the first sign of poverty and famine; a lack of nitrogen primarily in the form of protein quickly leads to a weakened condition and poor health. Nitrogen is a key element in essential molecules such as proteins, nucleic acids, enzymes, vitamins, and hormones. As we shall see later, world hunger is primarily protein hunger and, as the human population continues to grow, catastrophic famine.

With the aid of certain bacteria, a very small number of plants can convert, or fix N_2 directly into organic nitrogen, primarily in the form proteins. Most of the nitrogen in living organisms does not enter directly from the atmosphere. Instead, another group of bacteria in the soil and water can convert gaseous nitrogen, N_2, to inorganic nitrogen in the from of nitrate, or NO_3. Plants then obtain this nitrate as a nutrient from of nucleic acids and protein. Since nitrogen is also essential for photosynthesis, the amount of nitrate is the soil can regulate crop growth, thus explaining why the addition of artificial nitrate fertilizer can give greater yields.

When animals (herbivores)eat the plants, some of the organic nitrogen is transferred to these animals and eventually to other animals that feed upon these animals. When the plants and animals die, their organic nitrogen is then converted by decomposers to inorganic nitrogen in the form of ammonia, or NH_3 which is converted by another group of bacteria into another from of inorganic nitrogen, nitrite, NO_2. The ammonia, NH_3, can also be converted to atmospheric nitrogen, N_2, back to nitrate, NO_3, which can be taken up again by plants to begin the cycle again. Some nitrogen is lost from the cycle when soluble nitrates are washed from the soil and deposited as deep-sea sediments.

Disrupting the Nitrogen Cycle

Man's activities can now upset the nitrogen cycle. The danger here apparently is not in disrupting the cycle on a global scale but in upsetting it at the local level in a particular lake or soil system.

For example, consider what is happening to the natural nitrogen content in much of the soil in industrialized nations. Most of the nitrogen in the soil is contained in a complex mixture of organic and nitrogen cycle in the soil is functioning normally, very little nitrogen is lost, or washed, from the soil into nearby rivers or lakes and the fertility of the soil is maintained. However, if land is not planted properly or if it is over farmed, much of this natural soil nitrogen can be lost by surface runoff; a nutrient leak occurs and nitrogen can be lost by surface runoff; a nutrient leak occurs and nitrogen, primarily in the form of nitrate (NO_3) is transferred from the soil to nearby water systems. In addition to this leaching, or leakage, whenever there is harvesting, nitrogen is also removed. To maintain fertility it must be replaced by natural or artificial means. When we consider that it may take upto 500 years to build an inch of topsoil, we can appreciate the loss in fertility in American farms each year when over 2-billion tons of sediment are washed away by rain and irrigation.

As the land has been misused and slowly depleted of nutrients by continued use, we have been able to maintain or even increase crop yields by adding artificial, or man-made nitrogen in the from of inorganic fertilizers—as contrasted with organic fertilizers, proved naturally by waste and decay from plants and animals. Organic fertilizer cannot be used by plants until it is converted to inorganic nutrients by soil bacteria.

Unfortunately, overloading a soil with inorganic fertilizer apparently upsets the homeostatic balance of this crucial bacterial action. In the United States the use of inorganic chemical fertilizer has grown rapidly since World War II, increasing some 14-fold in the past 25 years. This has allowed us to grow more food, but at the same time it is causing some serious problems. Unfortunately,when inorganic nitrogen fertilizer is used, the amount of natural organic nitrogen in the soil can drop and the efficiency of the transfer of the nitrate from the soil to the plant can decrease. This means that some of the nitrate in the soil is not used and can be washed off into nearby streams or lost in other forms of nitrogen (for example, NH_3) to the atmosphere. Ironically, as a result of the extensive use of inorganic nitrogen fertilizer, a nitrogen nutrient leak has developed in many heavily farmed areas. The natural nitrogen content of the soil has been depleted while nearby water systems have become overloaded with excess nitrogen. This can result in a kind of "chemical vicious circle", where more and more fertilizer must be added as the soil becomes poorer and poorer and our water systems become more and more overburdened with too much nitrate.

What can excess nitrate or other nutrients (for example, phosphate) do to an aquatic ecosystem, such as a lake? Some chemicals and nutrient naturally run off of the land into a lake and as a result a lake becomes shallower over time as sediment builds up over the years. This natural aging process of nutrient enrichment and sedimentation is called eutrophication. All lakes naturally age and eventually "die". When man increases soil erosion or introduces excess quantities of sewage, industrial waste or fertilizer run-off, he can greatly accelerate the aging of the lake. This from of man-made eutrophication is called cultural eutrophication. The premature again and "partial death" of Lakh Erie is a good example of man's intervention in the natural chemical cycles. Again, we know relatively little about what happens in complex aquatic and soil ecosystems. Different lakes and soil systems have different characteristics and problems, but a crude general picture is emerging.

Plant life in a lake requires a balanced input of nitrates (and phosphates) as food, or nutrients. When plants or animals die or when organic matter or water is washed or dumped into a lake, as nitrate, phosphate, and other minerals that can be used to support plant and animal life. The lake is cleansed and life is maintained by the nitrogen cycle. This magnificent cleansing and nutrient cycle works well as long as it is not too heavily stressed. If the lake is overloaded by excess nitrates and phosphates, and entire nitrogen cycle and eventually the carbon and oxygen cycles can be disrupted on a local scale; with excess plant nutrient present, an explosive growth of plants, particularly algae, may occur. The resulting "algal blooms" cannot be supported for long and they soon die, overburdening the water with dead organic matter. The decomposing bacteria break down their dead matter by oxidation and in the process deplete the oxygen content of the water. If the process of over fertilization continues, the aquatic ecosystem has less ability to support fish and other aquatic life that require oxygen. As they die, this further overloads the system with excess dead organic matter, the oxygen content falls even further, the life sustaining and purification system collapses or is impaired, and the water becomes foul.

Unfortunately, in a modern industrialized society like the United States, there are many sources of nitrates and phosphates that can potentially overload many of our lakes. The "premature aging" of lake Erie may be merely an early warning system. Some of the major mancaused sources of nitrates and phosphates are (1) inorganic-fertilizer runoff, (2) municipal sewage-treatment plants, (3) animal-feed lot practices that produce massive amounts of animal wastes by concentrating thousands or hundreds of thousands of animals in a "livestock factory," (4) nitrogen-oxide emission from the international-combustion engine, and (5) phosphates found in most synthetic detergents. All of these can impinge simultaneously on a lake ecosystem to cause a massive nutrient overload and disruption of the natural rate of chemical cycling, as shown in Figure-3.

Obviously this process can be accelerated as the country increases its overall population along with increased industrialization and population concentration in areas around lakes (for example, the Great Lakes which contain 20 percent of all the fresh water on the earth). Furthermore, as population growing, more and more fertilizer must be used to maintain high yields of food and more and more animal wastes accumulate from our efficient feedlot system for grows, more livestock for the population. In the United States animal wastes produced each year amount to 2-billion tons—enough each year to fill a train reaching from there to the moon and halfway back and equivalent to the human waste produced by 2-million people. In other words, animals in the United States produce 10 time more waste than the entire US populations that we can have a meat-based diet. If the excess animal wastes were used as organic fertilizer, we could eliminate some of the need for inorganic fertilizer. Unfortunately, the problem at present is primarily economic rather than technological; inorganic fertilizer is relatively cheap. If we paid the real cost of inorganic fertilizers, including their devastating impact on our environment, then the use of organic fertilizer would again become economically feasible.

Ironically, as we spend more and more money in building primary sewage-treatment plants for our towns and cities, they can also add excess nitrates and phosphates to aquatic ecosystems and we will eventually be forced to go to much more expensive tertiary treatment plants that remove nitrates and phosphates.

The automobile, or more correctly the internal-combustion engine, is normally thought of in terms of air pollution, but it is also a contributor to nitrate water pollution. Large cities generally fall into one of two basic air-pollution classes the brown-air cities and the gray-air cities. Brown-air cities, like Los Angeles, are relatively young cities, where the main source of air pollution in the internal-(or infernal) combustion engine in the automobile. Every automobile spews several hundred different compounds into the air. One of these compounds is gaseous nitric oxide, NO, which is formed in the engine at high temperatures by the reaction of nitrogen, N_2 and Oxygen, O_2 in the air used to burn the fuel, Nitric oxide, NO, is relatively harmless but it can react with oxygen, O_2 in the air to form nitrogen dioxide, NO_2 a yellow-brown poisonous gas that produces the brown haze over car-oriented cities, such as Los Angeles.

Nitrogen dioxide may be harmful to the lungs. Even more serious is the fact that in the presence of sunlight it combines with carbon-hydrogen compounds (hydrocarbons), also emitted in auto exhausts, to form a whole new brew of chemicals. This mixture, called photo chemical smog, contains chemicals that irritate the eyes and damage vegetation. Furthermore the nitrogen dioxide can dissolve in water and be converted to nitrate, thus adding to the nitrate water-pollution to satisfactory level, it seems likely that the internal-combustion engine must be replaced.

Emission control devices now installed on cars do a poor job of reducing oxides of nitrogen and may actually increase their concentration. Many experts argue that emission control devices provide only a temporary solution, which buys us a little time (perhaps 5-10 year) to find a replacement for the internal-combustion engine. If, over a period of years, these devices reduce all dangerous air pollutants, on the average, by 50 per cent and during the same time the number of cars doubles (as projected), then air-pollution levels will return to present levels and even rise again as more cars are added.

The second type of city, the gray-air city, is an older city, such as New York, Chicago, and London, that depends heavily on the burning of coal or oil for heating, manufacturing, and electric-power generation. This burning normally produces two major air-pollutants: (1) particulates (dust and soot), which give the air over such cities its gray cast, and (2) sulfur-dioxide, one of the most dangerous pollutants in our air. Coal and soil usually are contaminated with sulfur, S, and when they are burned, the sulfur combines with the oxygen of the air to produce sulfur-dioxide gas, SO_2. The level of sulfur dioxide in our air, already critical, is expected to double in the next 20 years. It can damage plants, dissolve marble and concrete, and eat away iron and steel, not to mention what it does to delicate lung tissue. Exposure for 24 hours to only 0.2 parts per million of SO_2, with particulate matter also present in the air, is considered to be serious health hazard. The SO_2 is also connected to the nitrogen-dioxide problem from the automobile. SO_2 can react with oxygen, O_2, in the air to produce sulfur-trioxide gas, SO_3, which when breathed can react with water to produce droplets of sulfuric acid on lung tissue. Unfortunately, traces of nitrogen-dioxide, NO_2, in the air can serve as a catalyst to increase the rate of oxidation of SO_2 to SO_3.

Thus, can be begin to see that our air-pollution problem can be connected to another, and in some cases they in turn may be connected to a water-pollution problem. We can never do one thing to an ecosystem; the problem and solutions are complex, not simple. How can we minimize or reduce the over fertilization of lakes, such as Lake Erie? None really knows at this point. Obviously, reducing overall population and population concentration would help decrease inorganic fertilizer use, animal wastes, automobiles, detergents,and sewage-treatment plants. In addition, we might ban or reduce certain pollutants or use animal waste and municipal sewage as organic fertilizer instead of making more and more inorganic fertilizer as summarized in Figure-4. For example we might put research efforts on developing new technologies that could transport sewage and garbage to agricultural areas, thus completing rather than disrupting our natural ecological cycles.

We have seen that our life-support system depends on a magnificent system of chemical cycling and energy flow that developed during millions of years of evolution. We are now at a unique and unprecedented crossroads in the history of mankind. For the first time we have the capacity to destroy or seriously impair our life-support system, particularly at the localized level and global disruptions could occur within the next 100 to 150 years if we continue on our present path of more and more people consuming more and more resources.

A new dimension has been begun to appear in the last few years as the impact of man and his activities on nature came to be noted. The word 'environment' was given an additional meaning since about 1964 to express this broader concept of interaction of human activities on the ecological system and the impact of the ecosystem changes on man himself.

The impact of human activities on our physical and biological environment has a consequential effect on man and his society due to the changes in the environment. Illustrations of such changes occur in the archaeological record, such as the gradual desertification of parts of Sind and Rajasthan which one sustained larger populations and greater agricultural activity as well as more extensive grasslands. A dramatic evidence of such reciprocal interaction in our times is the discovery of mercury poisoning cases in some areas of Japan through eating of fish contaminated by mercury. For a few years the effect, or disease, in man was not even diagnosed as mercury poisoning but as a new illness. Later, it was traced to the mercury present in the fishes. Normally, mercury is only slight toxic. One can ingest a few grams of mercury without severe effects. However, research revealed that the high toxicity was due to poisonous compounds of mercury called methyl mercury (CH_2) Hg_2 and dimethyl mercury (CH_3) Hg_2. These were being produced by certain anaerobic micro organisms in the sea bottom of shallow areas of the east coast of Japan and were being ingested by fishes along with sea weeds which constitute their normal diet. The Physiology of fish was such that the poisonous mercury compounds tended to accumulate in some of the organs of the fish, the source of mercury was a product of industrial activity of pulp and paper mills which discharged small amounts of mercury in their effluents. The entire marine and human food chain is complex in which the micro-organisms of the sea bottom play an essential role to convert mercury and to link it with the food chains since the micro-organisms enter the marine food chain and work upwards to the edible marine fishes and ultimately to the human population who eat these fishes. Since methylated mercury are not easily eliminated by the human body, they tend to accumulate in the body, and over a period of time result in human illness and death. The diagnosis of the cause of the so-called Minamata disease and consequent preventive steps took some time as investigations over a period of years revealed the true nature and cause of the disease and the source of the pollutants.

The problem was not solved by discovery and prevention of the Minamata disease. Further researches in the Inland sea of Japan led to the interesting discovery of small quantities of methyl mercury in the fishes in the Kawasaki Bay close to the inland sea bed where there was no industry to cause mercury pollutants being dumped into the sea from industries anywhere near the area, further, the fishes in the inland sea were found generally to have lower methyl mercury concentrations. The source of mercury was finally traced to the volcanic pumice in the sea bed. Microorganisms produced methyl mercury and dimethyl mercury but nature's balance provides other microorganisms which break down the methyl and dimethyl mercury to mercury so that the concentration of the methyl and dimethyl mercury remains low. Recently the presence of mercury has been found in the fishes in the gulf on Cambay and in the coastal fishes in certain parts of the Bombay coast. The source of the mercury is still unidentified.

Less dramatic but, equally if not, more disastrous in the erosion of topsoil in the submontane slopes by floods and landslides preceded by and due to loss of vegetative cover resulting from tree felling (for fuel and logging) and loss in the process is estimated to be 105 mha in India. We are increasingly conscious of problems due to our paper mills effluents in many areas of the country, the acid sulphite process to separate the lignin from cellulose leads to formation of the lignosulphonates, these lignosulphonates are less biodegradable than the lignins and more toxic. Their persistence in water affects not only fish and aquatic fauna but also downstream users of water because of smell, toxicity, colour and decreased utility for human and agricultural use. Pesticides like DDT have been found in human beings in autopsies in Delhi and Bombay and in animals at several places because of their entry into the food chain in a variety of ways.

There are other well-known cases like the release of a toxic chemical (dioxin) from a chemical factory by an explosion in the factory in Seveso near Milan in Italy and its entry into the surrounding country and subsequently ingested by men and animals in the neighbourhood. More recently in Michigan a (polybrominated biphenyl) in bags normally used for packing entry of these compounds into the food chain through cattle feed affecting both cattle and human beings. It was first neurological disorders and skin diseases in man were traced to the presence of the chemical. In Delhi and Punjab, DDT has been found in cattle and human beings, children and young people tending to show greater amounts in their organs than older people.

There are many important activities of man apart from such accidents, industrial effluents and wastes, which have an impact on man or on the ecological system as a whole. The large scale burning of coal and petroleum as fuel for energy produces vast amounts of carbondioxide (CO_2). There has been a small but significant increase in CO_2 was 300 ppm (parts per million) around 1860 and is now in excess of about 400 ppm. It is thought that such increases, if continued, could result eventually in global, climatic and temperature changes, there are instances of rapid deforestations of hills slopes leading not only to reduced release of oxygen from plants by photosynthesis but also increased landslides and floods in the plains down below in U.P., Bihar and Assam. There is the encroaching desert in Rajasthan and Sind due to over grazing of marginal lands at the edge of desert, leading to the removal of green grassy cover of the land and decrease in the moisture retention capacity of the soil. This has resulted in decrease in rainfall. Consequently the land once marginally habitable has become desertified.

Technological activities of man have increased energy consumption, coal and petroleum are the main sources of energy today. In mining coal and extracting other minerals from under the earth, there is often subsidence of land and holes and gashes on the surface of the earth leading to the accumulation of water in these areas as well as changes of water courses and runoff affecting the ecology of these areas. It also gives rise to many environmental problems by local flooding, abandoned, water logged lands and increase in insect and water-borne diseases. The impact on the health of the community or of ugliness of the countryside is not always quantifiable.

The excessive use of fertilizers and pesticides contributes substantially to the ecological disturbances and environmental pollution. Excessive use of nitrogen fertilizers, for example, can lead to the leaching of these excess fertilizers into the water bodies.

These can be transformed by microorganisms into the nitrites and carcinogens such as nitrosoamines and can find their way back to men or animals through the food chain.

Excessive fertilizers can also adversely affect the nitrogen fixing organisms or microorganisms in the soil and reduce the natural nitrogen absorption and production of soils, similarly pesticides, besides killing the unwanted pests, insects or birds and can enter the human system through food or contact and become a health hazard. They can also indirectly come into the human system through animals, which ingest the plants. Other human activities also cause various ecological problems; for instance, over grazing in a grassland can denude it to the extent that it is gradually transformed into a desert. In Rajasthan, it is estimated that the desert in certain areas is advancing at the rate of half a kilometer a year.

Many of the large-scale human activities—industrial, agricultural or transport—tend to alter the natural balance of living organisms in an ecological system. Such a disturbance in the ecology of the system can affect human societies in a number of ways both in the immediate and in the long term. A grass land can be denuded and gradually transformed into a desert. Denudation of forest and grassland vegetation can cause long-term climate changes by reducing rainfall and thus perpetuate and strengthen the desertification process over decades and sometimes over centuries or millennia. The course of streams and rivers. It can also destroy the ability of trees, but the soil and humus which support underbrush, microorganisms, insects and animals which live in various symbiotic relationships with each other and the larger plants and trees. Once the forest is gone, these animals, plants and other organisms are also destroyed, making regeneration difficult or impossible. Larger trees and animals tend to be more valuable and are often the first to disappear.

Man is largely, if not wholly, dependent on his environment for the resources that sustain him and give him various products. In extracting these resources whether by way of agriculture or minerals or obtaining energy resources, the top soil may be over exploited leading to dust bowl-like conditions or so scarred as to be unfit for use. Forest may disappear if trees are felled faster than replacement. This leads not only to denudation of top soils but there may well be no fuel for the poor and no paper for the schools. One of the primary concerns of environmental management is the preservation of our natural resource base and wherever possible enhancing further its productivity in the long term. Unfortunately, the rapidly increasing population of our land has already doubled since independence. It is likely to exceed 100 crores by the end of the century. This large population needs food, energy, shelter and other amenities of life and in increasing amounts. The need of land is roughly proportional to the population if other things remain the same. However, the 150 million hectares of current agricultural land is not likely to increase very much. On to contrary, urbanization, development project, roads, etc. are going to encroach more and more on agricultural land and such land will be increasingly precious.

In various ways, therefore, the ecological systems and the human environment are closely connected. These linkages tend to confuse us sometimes, because the words are often used synoymously. Environment in our context will be used specifically from the human point of view. In some ways this leads to a larger concept which also takes into account human interactions with an ecological system and the mutual effects of human activities on nature and nature on man. When primitive man lived as a part of the ecological system, he was one of the many species of animals which interacted with one another and with the physical world. Civilized man has gradually, through the use of technology, transcended these symbiotic relationships to a considerable extent. This has given man a singular ability to interfere with or even damage nature, not always with wisdom or with advantage to his own long-term future. We have, however, become increasingly sensitive through the development of modern sciences of biology and ecology that ultimately man is extremely dependent on nature for his own survival and vulnerable to ecological changes. In a sense we have rediscovered the ancient wisdom of our land, the relation between Purusha and Prakriti'. If the ecological system is destroyed or seriously disturbed, this has the potentiality of jeopardizing human welfare, or if damage is great, of destroying human civilization. This recent awareness has, of course, been heightened by a persistent strain of philosophical awareness of an intimate relationship between man and nature in all societies.

Profound alterations of our environment are now being caused by the growth of modern technology, particularly during the last 100 years and at increasingly accelerating rates at present. Such alterations of our environment have been speeded up by the singular growth of human population at an almost exponential rate over the last 250 or 300 years, doubling almost every twenty-five years or thirty years, apart from population growth the other singular factor of our times is the many-pronged development of technology and the expectations that this has created in the process of techno-economic development of all societies including our country. The expectations of our people call for progressively greater availability of goods and services through processes of development. However, development requires utilization of our natural resources. If the use of these resources degrades the remaining resources for future use and depletes them in a manner that these are lost to future generations, we and our governments can be accused of being shortsighted, selfish and greedy. The activities of man are thus compounded by factors of increasing technological ability, energy and material consumption and large population increases. The environmental impact of such increased activities is large and is increasingly global in character. We should be concerned not only about the disturbances to the ecological system, but increasingly concerned seriously, in the long term, but the future of our planet. Man has now the capacity to destroy the entire living systems of the world for the first time in human history.

The close inter-dependencies between resources, development, population and environment seem obvious on examination. Unfortunately, quite often in the process of development, planning and the utilization of our resources we tend to ignore the factors of environmental risks and population pressures that are involved in the developmental process. There are other social factors, neither articulated clearly nor taken into account, such as discounted resource base estimation which given on framework for calculating what is happening to our resource base with time.

Today not only do all human activities, agricultural, transport or industrial, affect human perceptions of the environment but they also have their impact on the ecological system and the resource base. The four point focus of population, resources, development and environment gives rise to a complexity of relationships which both the development planner and the environmental manager has to take into account. However, this is not usually enough since human physiological and psychological factor such as health, aesthetic pleasure, scientific curiosity and technological skills are important for progress of a society for caring for the environment and building a sense of purpose and satisfaction in a society. Human perceptions include aesthetic perceptions and those of human health. Human beings and human societies have evolved a certain appreciation and communion with nature, forest and waterfall, sunsets and seashores have fascinated man from time immemorial. The preservation of these centres of communion with nature and of nature's beauty and harmony give great satisfaction to us. Such pleasure and satisfaction cannot be measured in economic terms or in terms of the technological achievement. Man likes, at least occasionally, to live in forest, hillsides and on seashores. Any society would be resentful if these are spoiled and rendered unusable. Natural beauty spots are easy to destroy and they are difficult and costly, and often impossible, to regenerate. The beauty and the harmony of nature that we perceive is almost always related to the harmony and stability that exist in an ecological system. The inner harmony of an ecological system is not always visible, the outward manifestation of its is often the perceived beauty and the sense of communion which has evolved as a concrete aspect of the impact of the ecological system on man is that of health. Man's physical as well as mental health appears to be connected to his symbiotic relationship with the ecological system in which he lives. Some scientists have theorized that disease is a symptom of maladjustment with the ecological system. Man certainly protected himself from disease in a variety of ways before historical times. These have become increasingly complex and sophisticated with our increasing knowledge of nature.

The concepts of our medical system have changed from that of treatment of diseases to their prevention. Prevention of disease is often thought of as the maintenance of an equilibrium relationship of man with his environment and not as a creation of completely lifeless or sterile surroundings for human beings. In fact it is neither desirable nor possible for human beings to live in a completely abiotic system. The concept of a relationship which enables man to remain healthy without forcing him live in a sterile atmosphere has strong support in our current knowledge of the living world.

As human population continues to increase in the next few decades, the need of adjusting ourselves to nature and of not disturbing ecological systems significantly will become even more necessary. The impact of human activities on the environment has to be taken into account and where necessary, ameliorated by suitable measures, so that human activities do not become self defeating in their not being able to create human well-being which was intended. Under increasing material and energy use, man's relationship with his environment will become exceedingly important and complex. Such use of materials and energy will usually lead to stresses in our environment. Societies and governments will have amount of social and self-discipline that will prevent serious stresses and distortions of the environment which might mean that a society would have to guard against degradation of ecological systems by groups of individuals, which either in the long term or even in the short term might affect human societies to their detriment. A methodology increasingly advocated at the national level is the laying down of a comprehensive environmental policy. Such a policy will necessarily have to be updated in the light of new knowledge and changing perceptions and values. The policy will have to be carried out in terms of actions and resources allocated for activities. Its execution starts with the education of children and adults and includes monitoring, investigation, research, enactment and enforcement of laws and encouragement of social and scientific groups which concern themselves with environmental questions. The total package of policy, law, education and activities is the task of environmental organizations. Environmental management can exist at several levels. Briefly, they can be institutional, industrial, civic, cooperative and governmental at local, state and federal levels, depending on the nature and size of the task of the issues.

Man increasingly lives in large agglomerations in the urban settlement or small ones in rural areas. We call them, in current vocabulary, human settlements. Human settlements are usually small and populations densities low in rural, agricultural, fishing or forest areas mainly because the amount of areas needed per person to sustain a population is larger. On the other hand,human beings needs each other for cultural, social and economic purposes and are thus drawn together to live in proximity. Human settlements, originally drawn together for economic, social and protective purposes created civilization and culture, philosophy and education, sanitation and medicine, communication and transport. As human settlements expanded these elements developed and assumed their current character. However, now more so than in the past, human settlements which are too small find difficulty in sustaining education, health and economic activities because there are not enough people to cover the wide range of activities that a modern society needs. On the other hand, beyond a certain population or forest areas are counter-productive because the land needed to support their needs of food, clothing and shelter increase beyond what the area can yield.

There tends to be an optimum population matched to a resource base and a state of technology above which poverty conditions begin increasingly to appear, similar arguments apply to the growth and limit of urban settlements which, although larger, also tend to be sustained not only by food, shelter and clothing but additionally by industry, trade, crafts, education, government activities, etc. Ancient urban centres, Thebes, Ayodhya, Athens, Rome or medieval cities of Delhi, London and Paris tended to grow but when they reached a certain population size depending on the state of technology and the resource are under the command of the city, they ceased to grow and sometime they declined.

The growth of modern technology and industry has increasingly helped to push up to population in urban settlements. In urban areas the dynamics of population increase under the impact of modern technology in the last two centuries have been to encourage large number of people to live close to industries and factories and to each other. In the last hundred years or less this trend has been served equally effectively by people trying themselves to transport systems. Industries and factories have tended to be created at convenient point where power, raw materials, transport and communications were easily available. This has led many industries to locate themselves close to each other, both for access to raw materials and transport and for the mutually supporting roles that industries tend to have with each other. The dynamics of industrialization and population increase have led to the growth of the large modern urban complexes. These have been extensively studied in recent years, the urban complexes are characterized by large per capita consumption of energy, the energy is required for the industries, for the transport of people and for providing food, domestic energy and other amenities to this large and closely knit population. The distinctive character, therefore, of urban complexes is high density of population, high concentration of energy consumption and a high quantity of wastes and effluents that both industries and the large population produce. While energy and material requirements for rural societies are at lower densities, transport, health and other requirements may have to be provided at equivalent or higher costs.

The two situations—urban and rural—are thus completely different. The urban settlement is sometimes called urban ecosystem, in which natural linkages have been significantly or largely destroyed or prevented and a new system of social and economic linkages have developed. The urban complex as a consequence has developed new sociological, psychological and technological parameters. The urban settlements are thus not only artificial in the sense that they are man-made, but they have also a different set of interactions and dynamics and do not fall strictly within the definition of ecological systems. Urban ecosystem is, however, an increasingly favorite terminology to describe the urban systems. The rural complexes on the other hand, still remain closely related to their environment and the linkages with the natural ecosystem are not entirely disrupted. While rural settlements also have problems, both psychological and technological in nature, these have yet to take acute forms. Part of the reason is perhaps the safety value in rural complexes tend to be the resource base, its optimal utilization, the various options on utilization,extension of the resource base, the impact of increasing population which the resource base has to sustain and the surplus it must create for urban centres.

Energy consumption per capita is comparatively low in our rural settlements consisting of cooking, drying, irrigation, etc. for rural life. The major source in most areas is energy from the surrounding biomass. This gives strong linkages to the ecosystem. On the other hand, the technology of agriculture and animal husbandry have changed substantially in recent years due to fertilizer and pesticide use to produce stresses and even changes in the ecological system. In a sense agricultural changes have further modified rural ecosystems in which new ecological relationships are being developed. Traditional agricultural, forest and tribal settlement systems have not escaped problems connected with intensive use of fertilizers and pesticides, detergents and plastics neighbouring areas, hence the management of agricultural wastes and residues, as well as those due to population growth and changing pattern of rural settlements are fast becoming matters of concern, these issues have been under study to some extent and while they pose what sometimes appear to be difficult problems, they are challenging and not necessarily intractable. Energy consumption per person in rural areas is currently small perhaps around 800 watt hours per day person. However, since a part of this energy has to transport in the form of fertilizers and pesticides as also in the form of electric power and fuel, the total energy consumption and distribution in the rural ares has to be taken into account to get an understandable picture. There is local generation of energy in the rural areas from agricultural wastes, fuel woods and to small extents by local generative systems such as wind-mills, solar energy and fuel forests, similarly nitrogen as fertilizer can be produced by culture of soil bacteria, legumes and associated rhizobia, and other methods. There are attempts in this direction in rice cultivation in Orissa. Phosphorus can be recycled through use of bone and fishmeal. These methods can, in theory, be so developed as to considerably reduce the energy needed to be imported from outside for the rural systems. In economic terms, the energy content imported is usually more than offset by the production of food and other agriculture-based products. There are however, environmental problems connected with the rural situation, particularly since the land and water, the primary resources, are limited. Distortions are often caused by wrong agricultural practices such as intensive monoculture of hybrids, massive use of pesticides and fertilizers and had water and soil management practices, leading to soil loss by erosion or salination.

There are other social aspects of rural areas. With the increasing local perceptions of deficiency of agricultural, educational, employment and cultural opportunities, rural dissatisfaction has increased and urban migration has accelerated. Such factors underline the perception that the quality of rural life is unsatisfactory. Trends of rural to urban migration have accelerated particularly as rural population increases have brought more pressure on a fixed amount of land tending to reduce per capita income. In such migration out of rural areas, there are both rural and urban problems. It is usually the adventurous,or the ambitious or the very poor amongst the rural youth who tend to migrate to urban areas, depriving a society of many of those who would be motivated and work for economic and social changes of the rural community.

Increase of urban population beyond the urban absorption capacities in terms of space, power, water, transport and handing of wastes of a city or a town on the other hand can lead to slums, social stresses and urban unemployment. The urban settlement is highly energy dependent and unlike the rural areas has very little or no local resources of eneregy. This high energy dependence leads to the question of the relation between energy consumption and the urban environment. The major sources of urban energy are coal,oil and hydroelectricity, although wood, dung cakes and straw are also used in the large towns, in the small towns, wood, dung cake as well as coal and oil are major sources. Minor sources of energy today are nuclear, wind and solar power of which wind and solar power are more appropriate to rural situations than urban. The combustion of coal and oil produces electrical energy or is used directly as energy for industry or transport. In all cases, it produces large amounts of carbon dioxide and particulates as well as various amounts of other gases such as carbon monoxide, sulphur dioxide, nitrogen oxides, gaseous, liquid and solid hydrocarbons, siliceous dust and mineral oxides. Hydroelectricity is often related to dam building and change of water courses leading to silting, destruction of valleys and salination of land. It also causes changes in the hydrology of the command areas. If the catchment areas are denuded of forest or vegetation cover, it can hasten silting of dams and cause landslides. Nuclear energy has problems of disposal of radioactive wastes and hazards of accidents such as melt down of reactor core which can spill radioactivity into adjacent areas, besides being costly and technology intensive.

We have thus a whole range of potential and actual environmental problems ranging from small perturbations in undisturbed ecological systems into which man is now making ingress such as introduction of new plants or animals into natural forests and natural lakes to entirely man-made human settlements such as cities and industrial complexes which are complex and artificial environmental systems. In man-made human settlements, man is in a sense primarily responsible for looking after the system to ensure and improve present human well-being. Man is also in a sense responsible to see that the resource base is not unduly jeopardized or badly exploited to the detriment of future people (our children and grandchildren) who will live and work in these human settlements.

In general, a good moral principle seems to be that present actions of a society should not jeopardize the future of the society or of any other neighbouring society. These considerations lead to the question of effective and good management of an environment, particularly when it is exploited for human benefit. There are many aspects—social, economic, legal, technological and biological involved in the management of such systems. The problem is complex and difficult, rendered more so, since data being incomplete, many factors are not usually fully known or understood. The problems are compounded in developing societies where the data and information base are poor, environmental awareness is lacking amongst the citizenry, pressures of rapid industrialization and urbanization are acute and there is lack of finances for environmental overheads. On the other hand, the need for taking care of the environment is often acute due to rapid deterioration of forests, agricultural lands, and waters.

Many general principles of management of any system apply equally to the management of environmental systems. However, the specific scientific, technological, social and biological aspects which relate to the environment, have also to be incorporated in the management of such systems. Some of these factors such as the effects of major engineering or technical activity may not be fully known or understood. They pose problems to which no ready answers are possible at the time that action is contemplated. There are aspects of uncertainty and risks which cannot be avoided when decisions have to be taken and action contemplated. There are many kinds of environmental systems and each has its own peculiarities and managerial problems. For most management purposes a broad classification of the various environmental systems is often adopted: (a) biosphere reserves, natural lakes or natural forests which are desired to be reserved for purposes of gene pool preservation, species protection, scientific studies such as those of evolutionary processes, aesthetic or entertainment purposes and the protection of geological or natural hydrological structures ;(b) rivers and river valleys, forests used as resources of timber and fuel, hill sides, beauty spots, etc; (c) agricultural land, agricultural waters, dams and artificial lakes; and (d) urban and industrial complexes and other highly disturbed zones such as mining areas.

It is generally assumed that all management is based on clear perceptions of objectives. Environmental management is equally dependent on clear formulation of the objectives in as specific a manner as possible both in social and ecological terms. Unfortunately, quite often, particularly in developing societies, such clear articulation of objectives may not be available due to lack of information on basic characteristics of the system, ignorance of impact of developmental activities, alternatives of objectives and consequential action, etc. In a biosphere reserve or a natural forest or lake, the problem of management may be stated simply that the objective is to have minimal human interference while scientific study and utilization to the extent necessary and possible is carried out cautiously, this means that the processes of natural evolution geological processes, energy and mass flows should be not only known and understood but also allowed to go on with as little interference, as possible. However, in most cases it is not necessarily, either possible, or desirable to fall back on a complete non-interference policy.

A legitimate questions would arise as to what are the responsibilities of management in case of a major flood or a cyclone, or an earthquake in a biosphere reserve or a natural forest lake or any other preserved ecological system.

We are aware generally of the serious ecological consequences of the great Assam earthquake and occasional major floods in some depth, similarly cyclones in the Orissa forest and coastal regions and the West Bengal and Bangladesh deltaic regions have been destructive of ecosystems. Although we are clear that we are responsible to take steps to ameliorate the situation, our ideas are still quite vague in such case as to what management can or should do both with regard to precautions and to actions after the event. We can and do take care of human lives in such disaster situations but elements of cost, ignorance and social organization make it difficult to enunciate policies and programmes related to animal and plant life.

Similarly, if we are dealing with the problem of a moderately disturbed ecosystem such as a river, the problems of fresh water fauna, the utilization of the river water by people who live near the banks further downstream, the problem of human settlements on the banks of the river at the location and downstream and the steps to prevent floods have to be looked at very carefully both in terms of aesthetic value of the natural river banks and the economic utilization of the river. The conflict inherent between maintaining the natural beauty of the two banks and the building of roads and parks or human habitation (including expensive hotels) on the banks will make competitive claims into which the environmental management system has to look closely. It has to approximate to achieving the clearly defined initial objectives with which it was entrusted. Social and economic pressures that have developed in the meantime will, in addition, have to be taken into account to prevent serious distortion.

As far as highly disturbed or rather the man-made environmental systems such as cities and farms, dams and roads are concerned, the problems are more acute. There is usually a greater clarity regarding what is desired even though this clarity may not apply equally to state how this is to be done. If a dam is built or a water reservoir has to be preserved one knows that within the precipitation basin of a dam or a reservoir, land erosion should be minimized so that the reservoir does not get filled up with silt in a short time. The problem often is how felling of trees or forests or other ground cover is to be regulated within the existing social parameters of energy needs and social habits of the people who previously depends on firewood or on the grazing needs of sheep, goats or cattle by nearby communities within the overall objectives of the projects.

In the case of industrial complexes, forest-based industries such as paper, mining and quarrying, the material resources which they will use and the disposal of their wastes and effluents have to be taken into account. The task is often clear if the system is well defined. However, the payoffs between the costs, social and economic and benefits such as the quality of life of people who work in the industry or of the people who live in the neighbourhood are not always quantifiable. Even more vague are calculations of benefit derived from a stable ecosystem in the long and short term.

Methods in applied ecology using simplified models continue to be developed to provide some theoretical base to understand the complexity of the system and suggest answers to policy questions such as possible fishing quotas or mesh size of fishing nets or spraying to control agricultural pests and insect-borne diseases such as malaria or use of biological agents for control of pests. The simple models do not provide a detailed and complete theoretical structure of ecosystems or even an understanding of the complexity of internal relationships. The availability of large computers have made it possible to incorporate increasingly complex and interlinked elements in a large system of equations to simulate more complex system for purposes of study. These cannot claim in any sense to be complete studies, but can in certain regimes or circumstances elucidate some understanding of the more complex behaviour of ecological systems.

While the formulation of applied ecological models began with the classic work of Lindemonn (1944) and has been pursued by Watt (1968), Jeffers (1972), Patten (1971, 1972) Van Dyne and Abransky (1973) and many others, there is a perceptible decrease in ambition in more recent work such as that of Austin and Cook (1974). Current aims are more to understand in general terms, the behaviour of ecological systems rather than to predict future behaviour of simple ecological systems such as succession or existence of multiple equalibria points or regions of instability or stability in systems, etc. A model ecological system can under constraints move from one equilibrium point to another, sometimes relatively easily. Unfortunately, such results are not quantitatively verifiable. Although there is evidence of ecological systems moving from one equilibrium point to another, such examples demonstrate the existence of limits of stability of ecological systems and of recovery and resiliency of such systems. However, theoretical exercises have not yet been successful in making detailed predictions of behaviour of ecological systems. The studies nevertheless are very useful and continue to be made because they do give indications of precautions to be taken or actions that should be modified or avoided in the course of developmental activities.

8 Land Management

Nothing seems more solid than a piece of land, and yet the plants and animals, the soil, and the life supporting nutrients provided by that land make up a single interdependent unit—an ecosystem that is dynamic on time scales ranging from days to seasons to years to millennia. Over days and seasons, the earth's plant communities absorb and release carbon in rhythmic manner. Over years and decades ecosystems respond to the natural patterns of plant succession and occasional events such as floods and drought.

Ecosystems function metabolically, producing and consuming many of the gases that drive the earth system. Plants capture energy from the sun and CO_2 from the atmosphere in their growth process. Terrestrial plants take up more than 100 billion metric tons of carbon each year and return approximately as much to the atmosphere as plants take more than 100 billion metric tons of carbon each year and return approximately as much to the atmosphere as plants die and decay. Micro-organisms in the soil release CO_2 and methane CH_4 as and products and nitrogen—containing tracegases as end products and nitrogen containing trace gases as by products.

Of the three main components of the earth system—atmosphere, water and land—the land is the most heterogeneous. The earth's surface is a mosaic of different types of ecosystems ranging from arid desert to tropical forests to tundra to more familiar temperate forests. Each harbors distinct plant and animal communities and each uniquely contributes to the functioning of earth-system. Land, being important to plants and animal communities is also the only component of ecosystem where man inhabits.

The total land area of the world is 13,128,841 hectares out of the total land use the crop land is 1,477,877 hectares, permanent pasture is 3,322,943 and forest and wood land is 4,095,317 hectares. Other land covers an area of 4,232,737 hectares and wild area is 3,486,097 hectares. The total land carries a population density of 403 (per 1,000 hectares).

The total land area of India is 32,87,263 Sq. Km., extending from the snow-covered Himalayan heights to tropical rainforests of the South. Out of the total land surface 43% is covered by plains, 28% is pateaus and less than 30% comprises mountains and hills. In all, only 62% of our total land area is topographically usable.

The three major components of our lands are croplands, forest lands and grazing lands—out of these grazing lands have been most neglected. Large areas of former grazing lands have come under agriculture leading to an overstocking of the remaining areas. The results is a hungry and malnourished animal population while experts debate whether the number of animals in India is not too large. Meanwhile, overgrazing prevents regeneration of forests and accelerates soil erosion, turning even foresters and environmentalists into enemies of the animal populations.

The environmental problems created by the neglect of India's grazing lands clearly shows that the current compartmentalised management of India's lands—croplands by departments of agriculture and irrigation, forests by departments of forests and grazing lands by department of animal husbandry, none of which talk to each other or care to understand each other's problems—can only lead to further deterioration of India's lands and improverishment of its rural populace.

India's lands are also being destroyed by indiscriminate mining activities with little ever being done to restore these lands. Displaced agriculturists and tribals undergo an overnight transformation from a relatively self-reliant people at the margins of subsistence to a callously abused and diseased mining community.

Grazing lands

India teams with animals. They come in all shapes and sizes, from the lumbering black buffalo to the ever nervous sheep and there are millions of them. India with just a fortieth of the total land area of the world supports more than half its buffaloes, 15 percent of its goats and 4 percent of its sheep. This livestock wealth plays a crucial role in Indian life, it is a major source of fuel and draught power, of nutrition, of raw materials for village industries. Even experts forget that Indian farmers are not just farmers; they practise a highly inter-dependent and integrated system of agriculture and animal husbandry.

Animal husbandry directly contributes 5 or 6 percent of India's national income but indirectly it makes much of the rest possible. As a source of manure and draught power, the cattle are worth their weight in gold for the millions of small farms. A number of village industries such as cane crushing and oil extraction would be virtually impossible without animals. There is also that 6 percent of India's population for which pastoral nomadism is a way of life. And, of course, without the study bullock cart, life in rural India would come to a halt.

But this huge mass of livestock needs fodder and there simply isn't enough of it. Only about 13 million hectares in the country are classified as permanent grazing lands. As these are obviously not enough, and exist in a highly degraded state, the animals scrounge for whatever they can get on millions of hectares of fallow and uncultivated land and uncultivable wastelands, as well as 36 million hectares of tropical forest lands. In all, almost all accessible vegetation in more than half the total land area of the country is grazed by livestock which snap up everything except non-palatable weeds.

Indian livestock appear to have adapted to the uncertain fodder situation. The animals have low rates of growth and production and can thus survive short periods of underfeeding and semi-starvation. They breed seasonally so the time when mother and calf need the most fodder coincides with goods grazing periods. Herds and flocks, particularly in drought-prone and desert areas and in the hills, migrate to places where prospects are better. However, the current fodder crisis cannot but aggravate difficulties when it comes to exploiting the full genetic potential of superior animals such as crossbreed cows, and is a major drag on efforts to develop breeds which offer more to society.

The effects of haphazard grazing on the environment are alarming. Land degradation due to overgrazing leads to desert like conditions which, in turn, reduce animal productivity and increase the economic pressure on human beings who depend on animals for their livelihood. Thus do pastoral nomads turn into landless labourers. And yet, of the three major land areas—grazing lands, forest lands and crop lands—

grazing lands have and continue to be the most neglected by the government.

One of the main reasons for overgrazing is too many animals and too little grass. There has been a progressive increase in all classes of cattle and buffaloes since 1951 and in 1987 there were as many as 180 million cattle and 62 million buffaloes, an increase attributed to the demand created by intensive agricultural practices. The number of goats alone increased from 47 million in 1951 to 68 million in 1982, despite the fact that the goat, a hardy creature which thrives on shrubs, bushes and thorns, has received no special treatment and that 36 percent of the goat population is slaughtered every year.

A second factor is that the government has just stood by and watched as grazing lands dwindled in area. in Gujarat, for instance, permanent pastures and grazing lands which spread over 1.04 million hectares in 1960-61 occupied only 852,000 hectares in 1986-87, while the bovine population had risen slightly; grazing land per animal was reduced from 0.11 hectare to 0.09 hectare over the same period. A large portion of Gujarat's grazing lands belong to the Panchyats and the powerful have encroached on them. In addition, there have been shift in the cropping pattern and a change in crop varieties. The area under fodder crops in Gujarat also declined from 8.16 percent in 1986-87.

In many cases, the government itself has launched programmes like irrigation schemes which have turned earlier grazing lands into croplands, but without any concern for rehabiliation of the graziers. The Rajasthan canal, for instance has brought vast tracts of lands under agriculture and the earlier nomads who used the area for grazing have been driven on to the shrinking grazing lands.

A third factor resulting in the excessive pressure on grazing lands is the absence of organised fodder production and pasture management. The growing human population has meant growing pressure on lands to produce foodgrains and raise cash crops. Forage crops do not find an important place in crop rotation because of small holdings.

Option for the future

The slow destruction of the country's grazing lands has created serious problems for the management of India's vast animal population. Particularly affected are the nomadic population. Experts meeting at the UN Conference on Desertification in Nairobi in 1977 agreed unanimously that nomadism and transhuman animal husbandry is the best possible productive system from an ecological point of view for the use of fragile ecosystems in arid and semi-arid areas.

The Government of India, however, does not appear to have accepted this wisdom. It has no real policy for helping nomadic groups adjust to the changing environment or to help them improve it. The general attitude of most government agencies is to devise ways to settle these groups. "The nomads in the present day society prove a menance for the whole society and their sedenterisation is inescapable", states the Department of Environment's "State of India's Environment" report. Government policies, ecological destruction and growing hostility from the settled populations have forced many nomadic groups to give up nomadism and become landless labourers.

Meanwhile, the natural balance that once existed between environmental resources and stocking rates is now being distributed. Environmental resources available to nomads have thus diminished noticeably ably because of inadequate government policies to protect grazing lands and active government policies to push crop production on marginal lands. To ensure their own survival livestock breeders have responded by increasing their animal populations and keeping more browsers. The result is heavier destruction of grazing lands and pressure on forests. The government, therefore, faces its biggest challenge in determining the future development strategy for its nomadic groups and in restoring the country's grazing lands to health. But politicians at all levels—from state chief ministers to Union cabinet ministers—have no understanding of the problems of nomads and would prefer them dead and gone.

In the past sporadic efforts have been made in different parts of the country to increase fodder production, by farmers on private lands. Work on breeding and economic aspects of fodder and pasture species was initiated at the Indian Agricultural Research Institute (IARI) at New Delhi in 1953. Research on fodder crops received further impetus with the establishment of the National Dairy Research Institute (NDRI) at Karnal in 1955 and the Central Arid Zone Research Institute (CAZRI) at Jodhpur in 1959. The Indian Grassland and Fodder Research Institute (IGFRI) was set up at Jhansi in 1962. A coordinated project on fodder crops was started in 1970 by ICAR with eight regional centres and six subcentres covering different agroclimatic regions. Agricultural universities located in different states also conduct some research on fodder crops.

Since the average holding is small, all the regional stations aim to incorporate fodder, particularly legumes, into existing crop rotations. Seeds of high-yielding varieties of fodder crops, now available for each region, are supplied by regional stations.

But almost nothing has been done about afforesting

public lands with fodder species and regenerating grasslands. Degraded grasslands contain a large proportion of non-edible shrubs. The IGFRI at Jhansi has found that the removal of the shrubs alone results in an increase in forage production of 1.2 tonnes to 4.2 tonnes per hectare per annum. Silvipastoral experiments have indicated that fodder-cum-fuel trees like leuccena, grasses and legumes can play a vital role in improving production of forage, fuel and grain.

Stall feeding has been repeatedly recommended by experts to stop ecological destruction caused by indiscriminate grazing, but this is clearly impossible unless there is a massive fodder production programme. Animal husbandry is a state subject. Fodder development is a law priority in state plans.

Immense prospects

The potential of grazing lands is immense in India. In Western Rajasthan, for instance, vast areas are available which are not suitable for agriculture. The total area classified as culturable wastelands and permanent pastures in the 11 districts of Western Rajasthan is estimated at 9.47 million hectares and 4.56 million hectares respectively. This area, even if moderately developed, can support a large livestock population.

The NCA, in 1976, had recommended that "green fodder production is the most important single factor on which will depend the success of the animal husbandry programme", that all unoccupied land should be planted with trees and bushes or seeded with self-regenerating grasses and legumes.

Social forestry programmes in arid and semi-arid areas should be based on fuel and fodder species. Ironically, government agencies and farmers are keen on a nonbrowsable species like eucalyptus specifically because animals do not eat its leaves. In all semi-arid areas, eucalyptus is now being planted. In many villages, social forestry programmes have been opposed because poor people (especially those with cattle but no land) are afraid that their grazing lands will diminish.

The Rajasthan forest department has started a programme for pasture land development through controlled grazing under the Desert Development Scheme. Since 1974, 311 village panchayats have been assisted in developing their grazing lands. The forest department plants suitable grass species and appoints a guard. The grassland is divided into two separate enclosures. After four to six months when the grass is big enough to be grazed, the department opens one enclosure and the sarpanch is told the number of livestock which can graze in a specified period. The guard is expected to regulate the grazing and collect the money from cattle owners, which is then given to the panchayat. This revenue covers the cost of production and maintenance of these pastures. The cost of maintenance comes to about Rs. 400 per hectare, which the panchayat pays, while the forest department develops the grazing land initially.

Dr. M.B. Mathur says, "This scheme has been functioning very well because of the villagers' full cooperation", This is despite the fact that pasture lands are closed for regeneration at a time when fodder scarcity is quite acute and opened after the monsoons from November to March.

A few voluntary agencies have also started taking an interest in plantations for fodder. The Maharashtra Dhangar Samaj Unnati Mandal in Kolhapur has been trying to involve local shepherds in planting *subabul* trees, if successful, this could be the first time that pasturalists who have usually been depended on the natural fodder base have fodder trees themselves. One of the oldest independent institutes which has worked for fodder trees is the Bhartiya Agro-Industries Foundation at Uruli Kanchan.

The answer to the huge problem of fodder lies in an appropriate and integrated national land-use plan. So far most attention has been paid to the area under cultivation. But nearly 50 percent of the land areas is uncultivated and most is uncultivable. India has not yet involved any policies or programmes for its unkeep and proper maintenance except stray, occasional and disconnected efforts. The hard fact is that management of forest lands, grazing lands and croplands can not be isolated from each other. The management of forest is dealt in separately. India being an agricultural land, it becomes important to know first about the soil, its structure and function, on which the crops grow.

Soil in Relation to Plant Development

The solid is the unconsolidated outer layer of the earth's crust, ranging in thickness from a mere film to somewhat more than 10 feet, which through processes of weathering and the incorporation of organic matter has become adapted to the growth of plant. It is underlaid usually by unconsolidated parent materials into which the deeper roots of plants frequently extend. The true soil or solum in generally made of the same parent materials which it covers. Nearly all higher plants except parasites and epiphytes are rooted in the soil. The latter do not depend on the soil directly but secure nutrients or support from plants thus anchored. Even floating water plants secure the necessary soluble salts that have been dissolved from the soil. The soil often contains and acts upon a much more extensive portion of the plant body than does the atmosphere. Moreover vegetation has played a remarkable role in the formation of this medium in which plants are anchored and from which they obtain their water and nutrients.

Nature and Origin of Soil—The geological formations, through weathering, produce the parent materials of soil. These materials constitute the bulk of the soil and for a long time determine its physical character. Thus the chief component of most soils is derived from rocks. More than 90 percent by weight of ordinary airdried soil consists of rock fragments. The small, angular fragments have been derived from the rock and possibly later transported by wind or water. During past centuries, rocks have disintegrated and they are still disintegrating by the action of such forces as alternate freezing and thawing, formation of ice in pores and crevices, erosion by wind and running water, and surface scouring by glaciers. The same physical forces that have weathered the rock have, in most cases, moved the fragments to other places where the soil is formed. Accompanying this process of fragmentation is the exceedingly important process of chemical corrosion or decomposition, for plants cannot grow in rock fragments, no matter how small the particles, unless the nutrients locked in these particles as insoluble compounds are changed chemically to water soluble substances. The latter alone can be absorbed by the roots. The fundamental reactions include hydrolysis, hydration, carbonation, and oxidation. They are common both to rock-weathering and to soil-forming processes.

The mineral matter derived from the rock material constitutes the matrix of the soil; it is very stable. The particles that have been exposed on or near the surface for centuries, even those of long-cultivated soils, scarcely differ from those of the deeper soil that have been protected and undisturbed.

The destructive physical and chemical activities of

weathering must be followed or accompanied by constructive biological forces in order to build a soil. Following or accompanying the accumulation of parent material there is the introduction of living matter which is largely responsible for the constructional processes of soil development. The residues of plants return more to the solid than the green plants have taken away. Throughout their lives plants have synthesized many organic substances such as sugars, starches, celluloses, fats, and proteins. Most of these materials return to the soil when the plant dies. The added organic matter produced by vegetation introduces a fundamental change. The substrate is no longer the former one of mineral matter alone, but now contains stored energy in the form of organic material. The soil soon becomes the abode of bacteria, fungi, and many other organisms. Throughout the whole process of soil development, plant and accompanying animal residues are converted, largely by the activities of micro-organisms, into the dark-colored organic matter of the soil. Thus, the two functions of living matter in soil development are synthesis and decomposition. The total organic matter or humus of a mineral soil constitutes from less than 1 to more than 15 percent of its dry weight. Sometimes soils are largely formed by the accumulations of organisms, as in peat or muck.

A true soil has five constituents: (1) the mineral particles of various sizes and in different stages of chemical decomposition; (2) organic matter in various stages of decomposition ranging from raw litter to well-decomposed humus: (3) the soil solution of various inorganic salts; (4) the soil atmosphere occupying interspaces not filled with soil solution; and (5) microorganisms, both plant and animal.

The chemical and physical charters of soils are

determined, at least for a time, by the kinds of rocks from which they are derived. Soil derived from sandstone, for example, has much coarser particles and, consequently, a lower water-holding capacity and better aeration that the clay derived from limestone. The latter, moreover, may be rich in calcium carbonate; the former has little or none. Clearly, these factors have a profound effect upon the vegetation growing on the two types of soil.

Soil Texture—Soils differ a great deal in the relative fineness or coarseness of the particles of which they are composed. These range from coarse gravel to dimensions below the visibility of the most powerful microscope. Texture is that property which has to do with the relative proportions of particles of different sizes. Of the three general groups of soil particles, the coarse ones are sand, those of medium size are silt, and the very fine particles are clay and colloids. The coarser particles of sand (0.2 millimeter in diameter or larger) have no colloidal properties and cannot absorb water or solutes. They do not cohere into crumbs or exhibit plasticity. Neither do they have any chemical action in the soil. They do have an important effect in counteracting certain unfavourable properties of clay.

The chief physical properties of clay are a high water-holding power, high plasticity or stickness, a property of swelling when moist and of shrinking when dry, often accompanied by soil cracking. In fact, most of the colloidal properties of the soil are due to the clay fraction and decomposed organic matter. The inorganic soil colloids are complex aluminosilicates resulting largely from the partial weathering of feldspars and other primary minerals. The nature and amount of solid colloids determine more than any other constituent the character of the soil. They are so tremendously important as retainers of soil water and nutrients and in furnishing these to root hairs and roots that they have been designated the "protoplasm of the soil".

Base exchange is an important chemical reaction occurring in both the inorganic and organic colloids of soil. The colloidal nucleus or micelle is negatively charged and has the power to adsorptions of calcium, magnesium, potassium, sodium, hydrogen, and other elements. Any cation can be replaced by another cation and thus go into solution and be made available for absorption by plant roots. Soil colloids are of three principal types which are designated as the hydrogen, the sodium, and the calcium colloids, which are more or less saturated with adsorbed positively charged hydrogen, sodium, or calcium ions, respectively. In the absence of calcium, magnesium, or other flocculating electrolytes, the hydrogen colloid is acid and easily dispersed into a collodal suspension which may be readily leached from the surface soil. The sodium colloid is highly alkaline and is even more easily dispersed and likewise leached. But the calcium colloid is mildly alkaline and remains in a flocculated condition. The colloids are never completely of one type in nature. Even the small absorbing power possessed by coarser soil particles may be due to a thin coating of collodial materials. Coarser soil particles are cemented by clay into soil crumbs. Soil without clay would resemble a pile of sand Chemically, clay is reactive in the presence of water. Fine sands and silts are intermediate in the presence of water. Fine sands and silts are intermediate in some of the properties described for coarse sands and clay.

The relative amounts of these different grades of particles in a soil determine its texture. The coarse soil materials, consisting mainly of quartz or shattered rock fragments, represent the "skeleton" of the soil. The function is largely that of anchoring the plant. Only the finer particles, the main products of weathering and decomposition, are very active in plant nutrition. The effect on the total surface of soil particles makes texture important as a soil property. The finer the particles the more surface is presented for the retention and solvent action of water and the greater is the absorbing area for plants. For example, coarse sand grains are 0.5 to 1 millimeter in diameter, but clay particles are so small that it requires approximately 65,000,000 of the largest of them to equal one grain of sand. The clay particles are so finely divided that they exhibit the properties of matter in the colloidal state, one of which is an exceedingly great water-retaining capacity. The surface area presented by a single cubic inch of clay is 10 or more square feet, that of coarse sand only about 100 square inches.

With respect to texture, soils may be grouped into a number of classes such as sand, sandy loam, clay loam, and clay. Each is determined by the proportion of silt, clay, and sand that it contains. The class to which a soil belongs can be closely approximated by examination in the field and determined accurately by mechanical analysis. This consists of separating the soil into the grades of particles of which it is composed and determining the percentage of each. The coarser sands are separated by means of screens, the fine Sand, silt, and clay, in water under the action of gravity after complete dispersion. This process of sedimentation is based on the principle that the larger particles sink more rapidly than the smaller ones.

The chief component of the first sample is silt, and because of its desirable proportions of sand and clay, the soil is designated as a clay loam. It has a good capacity for holding water—about 35 percent of its dry weight when undisturbed in the field. The second is fine sand with a relatively low power of water retention, only about 15 percent.

Many of the properties of a soil, such as content of air and water, depend upon the size of the soil particles rather than their composition. Soil texture not only exerts an important effect upon the water relations but also upon aeration, as well as the supply of nutrients. It profoundly affects the rapidity of the processes of decay of organic matter and its retention against leaching. The nitrogen content of a soil is closely related to its texture. A knowledge of soil texture is important in forestry, gardening, field-crop production, etc., and to an understanding of general plant distribution.

Soil Structure—Structure is a term expressing the arrangement of the individual grains and aggregates that make up the soil mass. The irregularity in size and shape of the rock particles prevents tight packing and affords open, irregular spaces through which air and water circulate, while their weight and mutual pressure furnish the necessary resistance for firm root anchorage. The soil is not a mere physical mixture of its component parts. The particles are intimately commingled in crumbs or aggregates cemented with the soil colloids. Soils of single-grain structure like Sand, where the particles function more or less separately are fairly simple. They occur where there are insufficient cementing colloids to bind the particles together. Sandy soils tend to be dry, loose and poor in soluble substances. A very complex structure is represented in clay where the soil granules or crumbs are composed of many particles, the size of the interspaces or pores is so small that neither water nor air can move freely. Such soils readily become waterlogged. In drying they shrink and crack badly, often damaging plant roots. During drought the water moves very slowly through clay, and plants may not receive the necessary supply rapidly enough. A rich loam usually furnishes an example of a soil with an excellent structure. Some of the particles are large and function as individuals. Those of smaller size form a nucleus about which the still smaller particles aggregate into granules, a process termed flocculation. This aggregation of the smallest soil particles into groups or crumbs which act as individuals makes the soil much more porous. The larger interspaces permit the water to drain away as they become filled with air, while the smaller ones retain moisture.

The increase of smaller solid particles somewhat retards the movements of both water and air. Hence, loam soils are characterized by a higher and more uniform water content throughout than are sands. The fine crumbly soil structure known as good tilth is best suited for plants. Soil in good tilth must possess a water-stable crumb structure. When the crumbs are small enough to permit of good aeration but large enough to bind the soil together an ideal state is reached. Weathering is the effective agent in crumb formation, but its action is greatly facilitated by proper cultivation. Humus has a very important effect in lightening a heavy soil and binding a loose one, thus promoting a good soil structure. Clay soils are usually improved in structure by the addition of lime which flocculates the particles into larger crumbs.

The structure of a soil largely determines its porosity. Thus, in turn, affects the absorption of water and, therefore, runoff and often the consequent erosion. Structure has a marked effect upon the susceptibility of soil to wind water erosion. During continued drought, soils without a cover of plants, even if they contain large amounts of clay, may be badly eroded by wind if the grains are grouped into fine crumbs or granules. Likewise the porosity of the underlying soil often determines the degree of erosion of bared soil on sloping lands. Where water penetrates too slowly, runoff and erosion are accelerated. Thus, soil structure affects the movement and storage of water in the soil and consequently soil aeration. Water content and aeration, as well as compactness of soil, all profoundly influence root development.

A good soil structure is maintained in nature by alternate wetting and drying, by freezing and thawing, by the action of organic matter and lime, and by the mechanical action of plants and burrowing animals Poor structure is produced by the puddling action of rain on bare soil surfaces, poor drainage, alkali, etc.

The importance of roots in maintaining a good structure is often overlooked. As a result of the interlacing and clutching of earth particles by myriads of roots, the soil is compressed into granules whose identity, stability, and permanence are established by a surrounding colloidal film of humified root material. Where the virgin prairie sod is first broken, it is mellow, moist, and rich. But after a few years of cultivation there occurs a great change in its physical condition. It becomes more compact, dries more quickly, bakes more readily, and often forms lumps and clods. But when grass is again grown for a few years, perfect tilth and freedom from clods are regained. The soil particles are wedged apart in places and crowded together in others. The small soil grains become aggregated together into larger ones. Each year many of the old roots die and are constantly replaced by new ones. The soil is filled with pores and old root channels; the humus from decaying vegetation helps cement soil particles into aggregates and thus lightens and enriches it.

The action of burrowing animals on soil structure is also important. Earthworms play a significant part. Their activities, in semiarid regions at least, are not confined to the surface layers. They sometimes penetrate to a depth of 10 feet. There are frequently thousands of them per acre. Burrowing everywhere, dragging down large vegetable fragments from above, they help to aerate the soil and keep it light. Rodents, ants, and various other animals mix and open up soil and subsoil and thus promote root penetration. On the "hard lands" of the Great plains, rodents have had an important effect upon soil structure, increasing water penetration and thus permitting the growth of certain deeply rooted species of plants. Insects, insect larvae, nematodes, and hosts of other organisms abound; all are instrumental in loosening the soil and thus affecting root development and plant growth.

Humus and Microorganisms

Humus comprises the more or less decayed organic portion of the soil. It consists of substances undergoing decay, of complex compounds resulting from decomposition, of constituents resistant to further decay, and of various substances synthesized by plant and animal microorganisms. It has no definite chemical composition since it is constantly changing as a result of the various processes of hydrolysis, oxidation, reduction, and condensation involved in decomposition. It has a high carbon content, usually about 55 percent, and usually about 3 to 6 percent nitrogen. In many soils, the elements carbon and nitrogen occur in the ratio of approximately 10 to 1. Humus comprises the total organic matter in soils. It originates almost entirely from decay of plants and microorganisms and, with rare exception, only a small amount is derived from feces or the remains of animals. Humus is dark in color, light in weight, and more or less intimately mixed with the other soil components. It does not dissolve in water, but combines readily with various inorganic soil constituents. It exists in the soil in the colloidal state with water as the medium of dispersion, the water being adsorbed by the humus. All soils that support vegetation contain humus. The amount may be very small in newly formed soils, such as windblown sand that bears only a sparse plant population, but it increases with the development of the sere.

Based on dry weight, the organic matter of a mineral soil may constitute from less than 1 to more than 15 percent of the soil. These low percentages are due to the fact that the mineral-soil matrix has a density about three times as great as that of the humus. Relative to volume, the humus may constitute 4 to 12 percent and the mineral-soil components only 41 to 62 percent, the remaining volume being pore space which is occupied by water and air. The amount of humus varies with the climate. Arid soils contain less, partly because there is less vegetation from which it may form, and partly because of its too rapid oxidation. For example, fine textured soils in Washington under 20 inches of rainfall have more than four times as much humus as soils of similar texture under 8 inches of precipitation. Moreover, in humid regions most of the humus is concentrated in the upper portion of the soil, the deeper soil being poorly supplied. In soils of semiarid and arid regions the organic matter is distributed more or less uniformly from near the surface to a considerable depth. In soils that are very wet, decay is greatly retarded and plant remains may accumulate in such quantities as to constitute 85 percent or more of the weight of the soil. This is the case in peat or muck.

Although much of the humus has its origin from above ground plant parts, large amounts are formed from root decay and a smaller amount from the remains of soil organisms. The decay of the organic debris is brought about almost entirely by the activities of various groups of bacteria, fungi, protozoa, and other inhabitants of the soil.

The microscopic plants and animals of the soil are

very diverse in food requirements but they are alike in that all finally depend upon decomposable organic matter. They are the universal scavengers, the organic matter regulating and determining the nature of microbial populations as well as their activities. They are most numerous in the surface 4 inches of rich, well aerated soil, in which the soil flora and fauna seem very similar the world over as regards species. During severe drought, the microorganisms live in the encysted form. Bacteria occur by the millions per gram of soil. They are most numerous in spring and fall, but their numbers vary from day to day and even from hour to hour. They usually recombine 5 to 10 percent of the carbon of the materials they have decomposed, but the amount varies widely. They thrive best in neutral soils and are mostly replaced by fungi in highly acid ones.

More than 250 species of soil-inhabiting fungi are known. Occurring in great abundance, they promote decomposition, often in soils that are too acid for bacteria to function effectively. Fungi transform into their own tissue 20 to 60 percent of the carbon of the compounds they have decomposed. By assimilating the ammonia and nitrates produced by decay, they save them from loss by leaching. Actinomycetes are a group of organisms, very abundant in soils, closely related to both bacteria and fungi. Many algae occur in the surface soils. Protoza, including amoebae ciliates, flagellates, etc., are of world-wide distribution. Certain species greatly decrease the bacterial populations by feeding upon them, but ultimately their remains are added to the humus of the soil. When all of the soil microorganisms are considered collectively, the amount of protoplasm in the surface 6 inches of a fertile soil is indeed great. Thus, significant amounts of micro-organic materials are added to the organic matter that these same organisms are so active in decomposing in their search for energy materials.

The first organic materials to disappear are the sugars, starches, pentosans, pectins, celluloses, and proteins. Their decomposition results in the synthesis of a proportional amount of microbial protoplasm. In fact, the nitrogen and mineral residues, such as phosphates, potassium, and calcium, made available by the decomposition of plant remains, may again be largely utilized by the microorganisms and released gradually upon their death and decay. Only a small part may at once be made available to the growing plant. This is fortunate, since otherwise the rapidity with which organic matter is decomposed would result in the liberation of nutrients which would be either largely lost to the atmosphere or leached beyond the roothorizon. Other organic materials such as lignins, fats, waxes, etc., are decomposed much more slowly by soil organisms. Thus, humus furnishes a small but continuous supply of nutrients for the growing plants.

In the presence of air, oxygen is slowly but continuously absorbed during the process of decay and an almost equal volume of carbon dioxide evolved. The organic materials are finally broken down into simpler compounds, the end products being carbon dioxide, water, ammonia, methane, and inorganic compounds of sulphur and phosphorus. With the decay of the tissues, compounds of calcium, potassium, manganese, etc., are likewise returned to the soil. The ammonia is immediately oxidized to nitrous and then to nitric compounds, which occur in the soil as nitrates. The process of decay, however, is not one of immediate simplification. Resynthesis of much material into microbial protoplasm occurs throughout. Moreover, the various organic acids, etc., originating as intermediary products, react upon the minerals with which they are in contact, and these may there by be made soluble and available to the plant.

Humus is being constantly formed and is continually being broken down. Although essentially a transient soil constituent, in an undisturbed soil it is fairly constant both in amount and in composition. Various stages in its decomposition are always in progress. It becomes an important source of nutrients for plants, and there is thus a close relation between the vegetation and the soil organisms. The latter are dependent almost entirely upon growing plants to furnish materials upon which they live, while vegetation is equally dependent on the activities of the soil organisms for removing the residues of previous generations plants and for the continued production in the soil of simple materials which are necessary to its growth. The enormous importance of this process in the economy of nature is evident, and the importance of the role of soil organisms can scarcely be overemphasized. Except for the work of the wrecking crews composed of myriads of microscopic organisms, which immediately attack the fallen vegetation and reduce it finally to water-soluble and, hence, usable compounds, plant residues would be distinctly detrimental to the soil.

Several factors profoundly influence the rate at which humus is formed and accumulates, as well as its chemical composition. These include (1) the nature of the plant and animal residues, including those or microorganisms, from which it is formed; (2) the nature of the microorganisms active in the processes of decomposition; and (3) the temperature, moisture and nitrogen supply, aeration, and soil reaction under which they are decomposed.

The decay of organic matter takes place most rapidly in warm, moist well aerated soil, i.e. under conditions most favourable to the development of the organisms carrying on these processes. In tropical climates there is little or no accumulation, despite the luxuriant vegetation. For each decrease of 10°C. in mean annual temperature, an increase of two to three times in humus content has been shown to occur. Raw humus is an accumulation of litter in cold or dry soils or those with an acid or alkaline reaction unfavourable to the growth of organisms causing decay. Three distinct layers occur in the raw humus of forests. Litter comprises the upper portion of the forest floor, slightly or not at all decomposed. The more or less decomposed organic matter just beneath the litter constitutes a layer of duff. A still deeper layer, in which the organic matter is so far decomposed that its original form is no longer distinguishable, is termed leaf mold. Certain coniferous forests and heaths of cool climates have soils of this type where the disintegration of the humus proceeds very slowly. Conversely, accumulation of litter in the deciduous forests, where it decomposes rapidly and adds its fertility to the soil, is small-seldom exceeding inch-compared with the northern coniferous forests. Where organisms causing decay are excluded because of poor aeration, the plant bodies accumulate as peat. The structures of the dead plants or parts of plants forming the peat are often preserved.

The effect of humus in improving the physical condition of the soil is marked. It acts as a weak cement to bind sand, lightens or opens a clay soil by separating the particles, and thus increases percolation, aeration, bacterial activity, and ease of root penetration. Being very absorbent, it helps to retain water so that in regions of moderate rainfall vegetation growing in soils rich in humus is less likely to suffer from drought. In fact, its physical effects are so marked that when the organic matter present in a soil is very high the distinctions between sands, loams, and clays are practically obliterated. Soils that have lost humus are harder than formerly and in poorer tilth; they crack easily and expose great surfaces to evaporation.

Development of Soil

Soils undergo a process of development somewhat analogous to that of vegetation. Both are a product of their environment. The controlling factors in the development of soil are climate and vegetation. Geologic materials including mantle-rock and bedrock are necessary, but are passive rather than active elements in soil develop through constructional process that involve the growth of vegetation, the accumulation and decomposition of organic substances, the formation of new chemical compounds, the progressive breaking down of the parent materials, the translocation of materials in the soil, and the formation of a more or less layered profile. Topography is important in so far as it affects the degree and kind of development through its influence on drainage and penetration and consequently on the water supply, salt content, amount of erosion, character and density of the vegetation, and other local conditions.

The development of soil and the production of vegetation are so intimately related that it is scarcely possible to study the one without some knowledge of the other. Functionally, the most important thing about a soil is its productivity for plants. Those characteristics which make a soil productive are themselves due to the operation of biological forces, particularly native vegetation under which the soil evolves.

The features assumed by the soil in its development from infancy through youth, maturity, and old age vary with the environment. Thus, all mature soils developed on undulating or gently sloping surfaces and for a long time undisturbed by injurious erosion, by deposit, or by the activities of man, owe their major characteristics not so much to the kind of rock from which they originated as to the nature of the climate in which they have developed. Hence, all stable, mature soils of a given climatic region belong to a climatic soil type. They tend to show a similar sequence of horizontal layers, irrespective of the underlying rock, the latter causing only minor differences. These layers occur in a definite sequence and differ from one another in one or more easily discernible features, such as color, lime content, texture, structure, and compaction. A vertical cut through these various horizons is termed a soil profile. The nature of the several layers has a profound effect upon the water, air, and nutrient relations of the soil and consequently upon root extent and distribution and nature of the plant cover. The parallel development of soil and vegetation and the role of each in modifying the other are of great interest and importance.

The constructional processes of soil development are due largely to the incorporation of plant parts among the mineral particles. This may occur following the accumulation of the parent materials or along with it. Vegetation introduces both directly and indirectly the biological factor into soil formation. Upon the fall of leaves and stems the organic matter of the plant, which has resulted from synthetic activity, is incorporated into the soil. These residues supply food for visible and microscopic soil organisms; these soil organisms decompose the plant remains, which ultimately become incorporated into the soil as humus.

The two most important kinds of vegetation thus active in soil formation are forests and grasslands. Differences of great significance occur among coniferous and deciduous trees and grasses as regards their rooting habits, the relative proportions and actual amounts of various elements they absorb, and the amounts they deposit in or on the soil surface with the fall of the leaves or the death of the tops. Organic remains under a forest of conifers have a much lower content of bases, such as calcium, than those under a deciduous forest. The ash (basic material) of a species varies within rather narrow limits, regardless of the nature of the soil on which it grows, but generally plants poor in bases and consequently producing an acid humus occur naturally only on acid soils. Usually the best growth of deciduous trees is made in areas having a good supply of bases in the soil or underlying parent materials. Under such conditions, the leaves are relatively high in bases, especially calcium. Grasses absorb, in general, large amounts of calcium and likewise return large amounts of this elements to the surface soil. Moreover, most grasses are only moderately deeply rooted and many species of arid climates are shallowly rooted. Where the bulk of absorption occurs in the surface layers, there is a tendency to reduce the amount of leaching through the upper part of the soil. The converse is true with deeply rooted vegetation.

Soil Profile

The rocks on weathering produce soil. At any place where parent material is weathering over a period of time, there develops layers of soil one over the other in a progressive state of maturity. Such a vertical section of soil from top mature soil to the rock below is called a soil profile. Not only do the layers differ in the size of their constituent particles, but very often they can be distinguished due to their colour differences.

In any soil profile the layers are named from top to bottom as A, B and C. Many ecologists distinguish another layer D. The true soil that is the weathered product of parent material is called solum and this constitutes horizons A and B. The semi broken region of the parent rock is called horizon C. In the A horizon the top fertile soil is richly mixed with organic material in different stages of disintegration. The top layer of fallen leaves and twigs still intact or only slightly broken is the Ao layer. Above it was the A0 layer of partly decomposed organic matter where decomposing organs have lost their identity. Fully decomposed organic matter i.e. humus enriches the soil by becoming mixed in with mineral components to form aggregations or crums. This is referred to as the Al layer and is dark in colour. A2 and A3 have a decreasing quantity of humus and are less dark than the A1 layer. In fact in warm humid tropical soil the distinction of the A horizon into so many sub-horizons is not clear because of the rapid biological activity in soil and the consequent quick humification and mineralization. The B horizon is also subdivided into B1, B2 and B3 with decreasing degree of maturity. B1 is similar to the A3 layer but some granular structures may be present, B2 and B3 horizons have soil black often formed due to aggregation of iron and aluminium. Horizon C represents a stage of weathered parent material which has not become true soil or solum.

Percolation and infiltration of rain water and its upward movement through capillary force to an actively evaporating surface leach down or raise up various materials and thus impart colour and chemical properties to different layers. Variation in climatic conditions and rock properties, therefore, lead to different types of profile and top soil development.

Zonal Soil Groups

Soils that are young, i.e. recently derived from rock or built up by deposit due to wind or water, as well as badly eroded soils, do not show a profile similar to that of mature soils upon which climate and vegetation have acted and reacted for long periods of time. The character of the climate eventually alters the nature of the soil from nearly all kinds of rock to such an extent that the mature soil type is similar in its main feature throughout a given climatic region.

The soils of the Great Plains, for example, consist of three great belts extending quite across the United States from Canada southward to Mexico. These correspond, in general, to differences in degree of grassland climate and in the type of vegetative cover. All of the soils (from near the Rocky Mountains eastward to about the one hundredth meridian), when mature, are dark in color and underlaid by a zone of lime-carbonate accumulation, beyond which water usually does not penetrate. They differ, however, in degree of darkness of soil color and depth of the carbonate layer. Where precipitation is least, the soil is brown and the carbonates have accumulated at depths of only 12 to 18 inches. This delimits the area in which water and nutrients are absorbed by the antive grasses. The effect of vegetation upon the profile is shown by its removal. Where land is broken, water penetrates quity beyond the carbonate layer. Had there been no vegetation to absorb the water, the layer of carbonate accumulation would probably never have developed.

Where precipitation increases eastward, the carbonate layer is deeper, vegetation is more abundant and taller, and the soil is of a dark brown color. Still farther eastward the carbonate accumulations usually reach depth of 5 to 6 feet. The soil is black in color and continues dark to a greather depth (8 to 24 inches) than in any other part of this climatic region. The vegetation is of the deeply rooted, mostly mixed-prairie type.

Another zonal soil type is represented under a

heavier rainfall eastward (prairie soils) where the soils are nearly black or dark brown in color and are reached or rock carbonates, which do not accumulate at a lower level. In this zonal type the soil is moist to a depth of many feet, in fact, often to the underlying water table. Roots of native plants penetrate very deeply (6 to 20 feet or more) and true prairie prevails.

Depending upon the nature of the climatic environment, there are three general processes of zonal soil development laterization-calcification and podzolization.

Laterite: This has been derived from the Latin word 'later' meaning brick. These soils are reddish brown in colour and are found in the warm and humid tropics. In laterization iron and aluminium oxides do not leach down but remain in the surface soil. Silicic acid on the other hand leaches to lower horizons. The Indian lateritic soils are, rich in Silica, Laterites are found in Australia, India, some warm parts of Europe and North America. These soils are comparatively less fertile.

Latosols: A wider term than laterite is now used more commonly to refer to soils of warm belt of the earth where silica usually leaches to lower horizons while on surface the oxides of iron and aluminium are present. These are poor in bases and have usually low organic content but rich in free quartz grains and kaolinite clays. The other names are oxisol and ferrasol. They thus become acidic as leaching continues. The soil fauna is depauperate. This soil is well developed in humid tropical or semitropical forested regions. Humus is quickly oxidized by the action of micro organisms and hence does not accumulate.

Vertisols: This is a characteristic tropical soil where there is a long dry season. These are also known as tropical black earths formed from limestones or ferromagnesium rocks and are rich in montmorillonite clays. Deep cracks are formed in dry summer. These cracks due to dustblow get filled with clay which in the rainy season on wetting swells so much as to form small bumps. Thus the vertisols are characterized by cracks in summer and bumps in rains.

Podozolic Soils: They are formed in northern, humid, cold temperate regions on heath lands and especially under coniferous forest. The condition thus necessary for Podozol formation are a humid climate where precipitation exceeds evaporation, an acid soil, and an acid humus producing flora. The acidic Podozol soils are very unfavourable to insect life earthworms, and bacteria. Low temperatures and poor aeration are contributing factors and the tannins from coniferous trees are thought to have a toxic action, especially to the bacteria. Bacteria are almost entire replaced by the more acid enduring fungi. Such soils are low in organic matter. Decomposition of organic matter by fungi leads to the production of organic acids and relatively soluble substances and to the reduction and solubility of iron. In normal soils, under conditions of good drainage, the soluble iron moves downwards, become oxidized and accumulates in B Horizon. A large proportion of the colloids both organic and inorganic, likewise becomes precipitated in this horizon. Thus, the surface soil becomes impoverished of fine colloids and organic matter and of nutrients most essential for the growth of crops. The leached soils are not fertile from the stand point of cultivated crops but, when clear of the native forest, are responsive to intense management and can support a successful agriculture of the subsistence type with emphasis on dairying.

The essential feature of a Podozol profile is its

division into three main horizons. A layer of organic matter (A0) consisting of leaves of trees and other forest debris, covers the surface to a depth varying from 1 to 12 inches. It is brown in colour, acid in reaction and only partly decayed. This raw humus layer may be sharply divided from the mineral soils below. The upper part of this (A1) is very thin and dark gray in color. The top varies from 'salt and pepper' grey to black depending upon the amount of incorporate organic matter. The lower part (A2 horizon) varies in thickness from a mere film to a foot or more and is light grey in colour. It has a friable or almost loose platy structure and has been thoroughly leached or eluviated.

The B horizon has a more compact consistency. It is reddish brown or nearly black in colour. The thickness varies from a mere film to several inches. The colour becomes lighter with depth, shading off gradually into that of the underlying parent material. This horizon has been enriched by the addition of eluminated organic matter, compounds of Iron and aluminium etc. The soil is acid in reaction.

Gray-brown Podozolic Soils: These soils have developed under a deciduous forest, rich in undergrowth, with a moist temperate climate. The surface horizons contain more nutrient elements and more organic matter. The undisturbed profile of the gray brown podozolic soils does not include a thick layer of raw humus. The accumulation of organic matter on the surface is rarely more than 1 or 2 inches in thickness. Although it is better decomposed than is the raw humus overlying well developed Podozols, yet it is not thoroughly decomposed into black, granular colloidal material.

The A1 horizon is a thin layer impregnated with

organic matter. It is thicker than corresponding horizon of Podozols and usually less acid in reaction. The A2 horizon is darker in color than that of Podozols being greyish brown or pale yellow. Neither is it usually so thoroughly eluviated. Like the B-horizon it is strongly acid. Compounds of iron and aluminum but not organic material have leached into the B- horizon.

Calcification : Is a common process found in North India. Here calcium in the form of calcium bicarbonate dissolves in water and leaches down escaping adsorption around dry particles. At depths of about half metre to three metres it precipitates depending upon the quantity of rainfall and depth of infiltration. The precipitated calcium remains is the form of nodules of Kanker and form a hard pan. This gradually reduces soil fertility. In Northern India the Kankar nodules are unearthed and are used in the building of roads and in the preparation of lime. Calcification is a common phenomenon in arid regions of temperate climates.

Chernozemic soil : These are found in the moist humid part of the drier regions having soils with a calcium carbonate horizon. This group of soils is most important in the world. The black soils of the Russian and Siberian steppes, the plains of both North and South America, and extensive areas in India and Australia belong to this group. They develop under a continental climate usually with a small excess of evaporation over precipitation. The luxuriant growth of grasses have produced black soils very high in content of organic matter and very fertile for cultivated cereals, especially wheat. Calcium and Magnesium are brought to the surface by the grasses. After their death and decay, the demands are united with carbonic acid to form their respective carbonates which are never leached beyond the depths of the roots of the grasses.

The deep distribution of humus through the profile is due in part to the action of earthworms, insects etc. and extensive burrowing of the rodents. In addition, a large part of the humus is derived from the roots of the plants which decay where they grew throughout the soil profile. Chernozems are not very productive not because of lack of innate fertility but owing to reduced precipitation.

Prairie Soils : These are grassland soils but are developed under a relatively high precipitation. They include the most humid upland grassland areas. They have been leached to an extent that they do not have a horizon of calcium carbonate. They are very dark brown in color, rich in organic matter, especially in the upper portion, and are well supplied with the nutrients necessary for a luxuriant growth of grasses. The surface soils over most of the area are slightly acidic. Their high fertility and the favourable climate make them among the most productive in the world for grain and grasses.

Gleization : Takes place in predominantly ice covered regions like arctics where soil is not saline. Peat accumulates on surface. Tundra soils are common example of gleization.

Tundra Soils : Soils of the arctic region are closely associated with bog soils because of the poor drainage. They are also connected with a climate sufficiently cold to prevent the slight summer than from extending deeply. They comprise the zonal group occurring in regions north of the Podozols. The vegetation of sedges lichens, mosses, etc. tend to form a surface layer of very slightly decomposed peat. Biological and chemical action is reduced to a minimum. Considerable mixing of mineral soil and peaty surface layer results from the expansion and contraction of water laden soil in consequence of alternate freezing and thawing. As the surface layer freeze pressures exerted on viscous mass beneath forces it upward through the cracks. The characteristic blue grey layer of slightly weathered clays, silts and sands beneath the peat is indicative of poor aeration.

Intrazonal Soils : Saline soils (Solon chak) alkali soil (Solonetz) and degraded alkali soils (Soloth) are local intrazonal soil groups and are not zonal soil types. They occur in regions that are receiving or have received at some time, an excess of salts. They are areas of periodic excessive moistening and drying. Bog soils (organic soils) are where peaty materials have accumulated under the influence of excessive moistening and deficient aeration.

Desertic Soil : Are characteristic of arid climates and contain very little organic matter. A profile is poorly developed. The surface soil is brownish grey and grades quickly into calcium carbonate horizon which usually forms a hard pan just below the surface. Wind erosion removes the finer particles leaving, the coarser material to form a hard pavement. The soils are but slightly weathered and leached, lacking nitrogen, and infertile.

Mountains and Mountain Valley soils: They vary from shallow layer of eroding rocks to deep organic soils of valleys and swampy areas.

Alluvial Soils are deposited by water current along the course of rivers. They are usually very fertile and support luxuriant vegetation. They are mostly without a developed profile.

Land Degradation

The rapid industrialization, developing of new cities, urbanization, clearing of forests, and other human activities have imposed a negative impact on our land resources. Estimates of land damaged or lost for agriculture use through soil degradation range from moderate to apocalyptic. The types of degradation are as land pressures in rural areas. Three aspects of soils degradation—desertification, erosion and salinization or water logging—receive the most attention although desertification does not have as large and pervasive an effect on productivity as do others. Desertification in the form of advancing frontiers of sand that engulf pastures and agricultural land is not the most serious problem in dryland areas, although it occurs locally. Definition of desertification, however are usually broader and include losses of vegetative cover and plant diversity that are attributable in some part to human activity, as well as the element of irreversibility. Desertification in this sense is difficult to measure. It is clearly affecting some dryland regions, but truly irreversible damage is probably less widespread than is commonly believed.

More widespread than desertification, if less dramatic, is the gradual deterioration of agricultural soils, particularly in dryland areas. Results of a global assessment of soil degradation sponsored by U.N.E.P. show that 1.2 billion hectares—almost 11% of the earth's vegetated surface have undergone moderate or worse soil degradation over the past forty five years because of human activity. As a result of this deterioration yields and total harvests of important crops are declining in a number of countries. Erosion is one of the key components of soil degradation. Its irreversibility and its potential offsite effects distinguish it from the other critical elements of soil deterioration—loss of plant nutrients, organic matter, and micro-organisms. Soil erosion is actually removal or erosion of top layer of the soil hence involving loss of fertile layers of the soil. Its main causes are high velocity wind, heavy rains.

As rain water drains over the surface of the ground,

it picks up and transports particules of soil more or less proportional to the rate of flow. This is a continuous process that has gone on throughout geological time, resulting in base-leveling of mountains, the filling in of valleys, and the formation of river-deltas. Normally this is a slow process, with changes becoming apparent only in thousand years. With the ground covered with vegetation, the flow of water is retarded and more of it percolates into the ground or is evaporated. In fertile areas, new soil is formed faster than it is eroded or at least there is stabilization.

With removal of the native vegetation and cultivation of the soil in agriculture, surface particles are more easily moved and surface water drainage is more rapid. As a result, top soil is being lost at an alarming rate. Much land has been so badly damaged that there is little likelihood that it will ever again be productive crops. Conservation of the fertile top soil is one of major problems facing modern agriculture. Putting in underground drainage tiles in level farmlands helps to draw rain water into the soil and to decrease surface run off and erosion. Soil coservation procedures should be strictly followed.

Agronomic research and project experience are that erosion is best prevented through balanced management of soil moisture nutrients and organic matter. Low cost techniques for soil conservation, designed to improve soil moisture levels, can increase yields sufficiently within the first several years to make the interventions profitable in their own right, regardless of the long run benefits of soil conservation. Compared with traditional cropping methods, practices such as mulching, manuring, low tillage, contour cultivation and agroforestry can frequently reduce surface run off water sediment loss and erosion by 50% and more. These techniques are not yet widely used. They will contribute to the control of soil degradation only if practical constraints such as shortage of cash and labour and the use of dung and mulching materials as household fuel are first alleviated. With the rapid expansion of irrigation over the past forty years have come growing problems with salinization and water logging that are eating away at the productivity of irrigation investments. Irrigated land in deteriorating in parts of many countries including China, Egypt, India, Mexico, Pakistan, the Central Asian Republics and Western United States.

Salinization or irrigated land is part of a much larger problem of managing the productivity of soils affected by salts. Globally perhaps about 950 million hectares, or nearly one-third of arable land, are affected by elevated salt concentrations. Most of this salinization occurs naturally. But about 60 million hectares, or some 24% of all irrigated land, suffer from salinization caused by bad irrigation practices. Severe declines in productivity affect, according to some estimates, about 24 million hectares or about one tenth of irrigated land. Despite awareness of these problems, and despite several decades of reclamation efforts, new areas are being degraded faster than other soils are being rehabilitated. Prevention and reclamation may continue to be hampered by the cost and managerial complexity involved.

INDEX